Cost Control in Building Design

Cost Control in Building Design

An interactive learning text

Roger Flanagan
Department of Construction Management & Engineering
The University of Reading

and

Brian Tate
Department of Land and Construction Management
University of Portsmouth

Blackwell
Science

Editorial Offices:
Osney Mead, Oxford OX2 0EL
25 John Street, London WC1N 2BL
23 Ainslie Place, Edinburgh EH3 6AJ
350 Main Street, Malden
MA 02148 5018, USA
54 University Street, Carlton
Victoria 3053, Australia

Other Editorial Offices:

Blackwell Wissenschafts-Verlag GmbH
Kurfürstendamm 57
10707 Berlin, Germany

Zehetnergasse 6
A-1140 Wien
Austria

First published 1997

Set in 10/13 Zapf
by DP Photosetting, Aylesbury, Bucks
Printed and bound in Great Britain by
Hartnolls Ltd, Bodmin, Cornwall

The Blackwell Science logo is a trade mark of Blackwell Science Ltd, registered at the United Kingdom Trade Marks Registry

DISTRIBUTORS

Marston Book Services Ltd
PO Box 269
Abingdon
Oxon OX14 4YN
(*Orders:* Tel: 01235 465500
Fax: 01235 465555)

USA
Blackwell Science, Inc.
Commerce Place
350 Main Street
Malden, MA 02148 5018
(*Orders:* Tel: 800 759 6102
617 388 8250
Fax: 617 388 8255)

Canada
Copp Clark Professional
200 Adelaide Street, West, 3rd Floor
Toronto, Ontario M5H 1W7
(*Orders:* Tel: 416 597-1616
800 815 9417
Fax: 416 597 1617)

Australia
Blackwell Science Pty Ltd
54 University Street
Carlton, Victoria 3053
(*Orders:* Tel: 03 9347 0300
Fax: 03 9347 5001)

A catalogue record for this title is available from the British Library

ISBN 0-632-04028-9

Library of Congress
Cataloging-in-Publication Data
is available

Contents

Preface

The importance of cost control in the construction industry does not have to be justified. Clients want projects to be built within budget, on time, and to their required standard. Contractors and specialist contractors want to build a facility to meet the clients' needs within the tender figure, but also ensures they make a reasonable profit. Cost control is at the heart of the management of the construction process.

This book is notable for two reasons. Firstly, it explores all the issues that affect cost control on a construction project. Secondly, the contents are presented in the form of a 'programmed text'. It uses as its starting point the HMSO programmed text on cost control in building design produced in the 1960s. Over the past 30 years a great deal has changed in the construction industry. There are now severe pressures to reduce the cost of projects and to build faster. Cost cannot simply be monitored, it must be managed and controlled. That requires a good understanding of cost planning from the concept stage right through to completion.

There is a great deal for the student and young professional to learn. The purpose of this book is to develop the ideas of cost control from first principles and to take them forward into practice.

Programmed learning has shown its potential in many spheres. The main advantages are that individuals can go at the pace that is best for them. The material is presented in a readily understandable form which can be followed at any convenient time and is not restricted to a set of lectures or talks.

The material in the book is equally applicable to any project in the international construction market place. Cost control and cost planning are as appropriate to projects being built in the developing world as they are in the developed world. This book uses examples from the British construction industry, but is not restricted to simply British thinking.

The book is divided into two parts, the first part deals with the principles of cost control. The second part deals with the techniques of cost control. Worked examples are used throughout to guide the reader from first principles to the most up-to-date thinking in the area of cost control and management.

Part 1

The Principles of Cost Control

1 The Importance of Control over Expenditure

This chapter *is not* in programmed form.

The construction industry and the environment in which it operates have changed significantly – the process of change now seems to be never ending and hectic. There have been:

- new approaches to buying design and construction
- fragmentation of the industry with the increase in specialist trade contracting
- the demise of direct employment and the growth of labour-only and fixed-term contracts
- fee competition for consultancy services
- the introduction of partnering
- more conflict and less trust
- clients wanting more value for money.

The boom–bust business cycles mean that the industry lurches from over-supply of resources to under supply. Against this background, contractors and consultants have to cope with the ever-increasing pressures for faster construction to a higher quality at a lower cost. Information technology has also made an impact with computer aided design (CAD) and electronic data interchange of information.

However, what has not changed is the importance of effective cost control. Costs should not be simply monitored, they need to be controlled and managed from the early design stage through to project completion. Some CAD systems give cost information, but they cannot control costs, that requires specialist human knowledge.

Clients want:

- certainty of price
- projects constructed within budget

- completion on time
- the best quality possible for the price
- value for money
- no surprises.

Contractors and consultants want:

- a reasonable return for the risks they take
- payment on time
- clients who do not keep changing their minds
- satisfied clients
- repeat business.

The aims of clients and contractors are not in conflict; they are complementary.

Four pressures

In this chapter we examine how change has come about by reference to four pressures:

1 *Society is having to cope with rapid technological and sociological change at a pace never seen before. Managing risk, avoiding unpleasant surprises ensuring value for money and speeding up the overall project delivery time is important for clients.*

'Time is money' said Samuel Johnson, and it is still true today. Speed of design and construction have become very important. The traditional route to getting something built was to fully design the project, produce a bill of quantities, obtain tenders, award the contract to the lowest bidder. The system is tried and tested and is still largely used today. The failing is that the project is often only half or three-quarters designed when it goes to tender. The shortcoming is not the fault of the design team, it is simply that pressures cause the project to get to the construction phase as quickly as possible.

Procurement systems which overlap the design and construction phase have become accepted practice for larger projects. Management contracting, construction management, and various other agency fee arrangements adopt this approach. Here cost control is exercised via a cost plan based on work packages where both design and production costs jointly come under scrutiny.

Design and build, sometimes called package deal and turnkey approaches, can also be used by clients where they want single-point responsibility. However, even here clients will demand early cost advice and budget control from their professional advisers, before finally passing the risk for cost control to the contractor. The degree and method of cost control exercised by consultants will depend on the type of design and build used.

For some, the traditional approach is more appropriate, where design consultants produce a completed or near complete design prior to start on site. This will entail a fairly lengthy design stage and consideration needs to be given to controlling costs over this stage.

One way or another, both in the commercial and non-commercial world, time is important to many clients. Where design is separated from the production stage, clients are less willing to tolerate the delay caused by redesigning buildings when tenders are too high. We therefore need a technique for producing designs for buildings which can be built within the expenditure limits.

When the risk of cost control is passed to the contractor, as with most forms of design and build contracts, then the contractor will develop suitable cost control strategies. Although much of the design might be complete prior to start on site, nevertheless the design still has a certain amount of flexibility built into it. Thus design and production costs are brought together under the control of the contractor. Given this situation it is likely that the contractor will use cost control procedures which look at resource costs. Resource costing is a subject in its own right which is only briefly discussed later.

2 *The second pressure is that construction projects are more complex. This has been brought about, in part, by the requirements of clients who know what they want to achieve. In brief, clients' requirements are becoming more complicated and more demanding.*

Wherever you look, buildings and project schemes are becoming more complex. Take as an example a typical modern shopping centre located in the heart of a city, with all shopping units housed under one roof, providing covered malls, integrated parking facilities, and goods storage, but separating the means of goods delivery. Consider the control of the internal environment, fire safety, security control systems, and the trading requirements with point of sale computer systems. This is very different to constructing the one-off shop in the village street. Similarly, think of a hospital project where 50% of the cost is in the environmental engineering services.

Much of the complexity comes about because of the environmental engineering services element. More thought is given to the internal environment. For example, atria are now commonly incorporated into building designs. Another example is the increased use of information technology which has meant that buildings designed no more than 20 years ago are technologically obsolete.

This increasing technical complexity of construction means that there are many more opportunities for the cost of a project to get out of hand. We therefore need an effective system for the strict control of costs from inception to completion of a project and then onwards throughout the life of the facility. Throughout this programme we will be discussing control of costs. Unless otherwise stated, 'cost' shall mean 'cost to the client'.

3 *The third pressure stems from the increased number of groups who have interests in a project.*

Undertaking a construction project is a complex process which involves many bodies and organisations. Effective co-ordination between the various parties is crucial for successful completion of both design and construction. In general, these parties are frequently referred to as the 'client', the 'design team' and the 'construction group'.

The 'client' can be an owner-occupier, a pension fund, an insurance company, a development company, a retail organisation, a large industrial corporation, a privatised utility, a housing association, or a government department. The full list is extensive. Some build for profit (office blocks), some as a factor of production (factories, oil rigs), some to fulfil a social need (old people's homes, schools), and some to enjoy the use of the facility (churches, housing). Most importantly, the word 'client' (referred to as the owner or promoter in many countries) is a catch-all term that hides a multitude of interests and needs. All clients are different in their expectations, and the design and construction team must recognise this. However, in every case the management of the cost will be high on their agenda.

Client group

The client group will frequently comprise a range of members including:

- ❑ technical advisers
- ❑ legal advisers
- ❑ financial advisers
- ❑ property advisers
- ❑ maintenance engineers
- ❑ project managers
- ❑ manufacturing/process experts.

Design team

Design teams will frequently comprise:

- ❑ architects
- ❑ structural engineers
- ❑ quantity surveyors/cost consultants
- ❑ project managers
- ❑ acoustics engineers
- ❑ interior designers
- ❑ landscape architects
- ❑ hydraulics engineers
- ❑ mechanical electrical engineers.

Construction group

The construction group will include:

- ❑ the general contractor
- ❑ specialist contractors
- ❑ specialist suppliers
- ❑ component manufacturers
- ❑ sub-contractors
- ❑ utility companies including gas, electricity, water, telephone
- ❑ building materials suppliers.

Other groups

Other groups who may have an interest in the construction process include:

- ❑ central government
- ❑ local authorities (planning, highways and building control)
- ❑ labour unions
- ❑ heritage/conservation groups.

The funding of construction projects has become very complex. For example, on the Channel Tunnel project there were 220 financial institutions involved. On a commercial development project built to lease to tenants there could be many funding bodies, ranging from the developer to, not least, the final occupier or tenant. A project might have an international flavour with overseas banks involved, or there might be a grant from the European Union or similar body.

There are few projects in the world where the design team has been told not to worry about the cost!

Public sector organisations are subject to audit and public scrutiny of their accounts; value for money and accountability are high on their priority list. In the private sector, a public company must act in the best interests of the shareholders, while a privately owned company will always be seeking to optimise expenditure. In every case, whenever an organisation gets involved in a construction project, the overriding objective will be to manage the cost within budget.

Actual building expenditure must accord closely to estimated expenditure. The cost consultants are usually instructed to forecast and budget the cost of a building at the very early concept stage before drawings and specifications are available, and they are often required to keep to their first estimate. Clients are increasingly resorting to litigation when things go wrong.

4 *The fourth pressure stems from modern practice in design, where new ideas, techniques, materials and components are used.*

Methods of construction or, perhaps more accurately, of assembly have advanced. Improvements in site plant and equipment, both large and small, have transformed the site processes. Architectural ideas have changed and designs have progressed, so much so that at times there is no precedent or data base from previous projects upon which to base estimates.

This increasing range of choice in materials, types of building and architectural developments in design styles can have a tendency to make one-off initial cost estimates unreliable so that the final account is seldom equal to this first estimate. Under such conditions continuous cost control is required.

Where continuous cost control has not been exercised, then it is quite likely that tenders received will exceed a client's budget limit. In such a situation the usual response is to look for savings within the specification. At this late stage, such action usually has undesirable implications for the architect's carefully thought-out design. Accordingly, unless effective techniques are put into place from the outset, it might not be possible to achieve a balanced design and thus ensure value for money.

The cost planning framework

Cost planning is a system of relating the design of buildings to their cost, so that, while taking full account of quality, utility and appearance, the cost is planned to be within the economic limit of expenditure. Cost planning procedures are applied in an attempt to reduce the amount of resources (and therefore cost) incurred during each stage of the development process, including design, construction, operation and maintenance, and subsequent replacement.

Although effective execution of the development process will demand an integrated approach to cost planning, this book focuses solely on the design process. Design cost planning is particularly crucial as decisions made during the early stages of the development process carry more far-reaching economic consequences than the relatively limited decisions which can be made later in the process, as illustrated by Fig. 1.1.

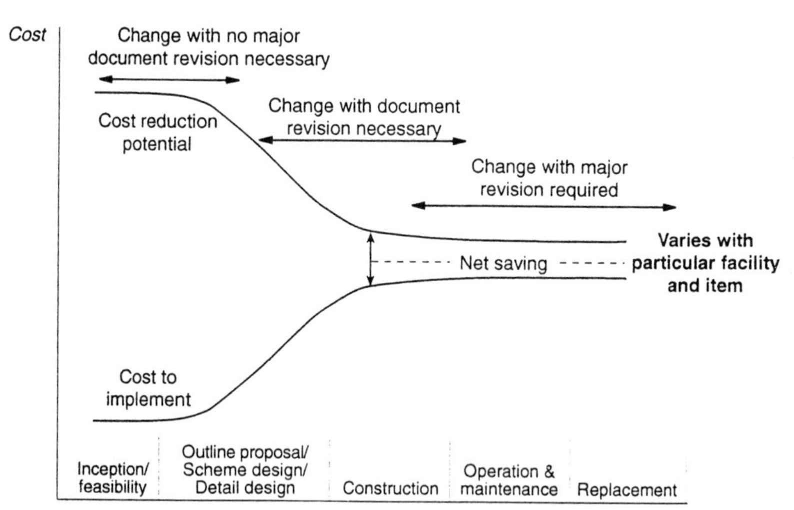

Fig. 1.1 Impact on cost of incorporating a corrective change or a better idea

Figure 1.1 shows the typical life cycle of a project from concept through to operation and occupation. The cost reduction potential curve illustrates that the most significant cost reductions are achieved during the design and development stages of the project. During these stages large cost savings can be achieved as the design is flexible enough to incorporate relatively significant changes. Once the project has reached the construction stage, the potential for achieving cost reductions is significantly lower. Additionally, any cost reduction associated with this curve will have a resultant cost of implementation. The 'cost to implement' curve must be viewed against the 'cost reduction' curve in order to assess the net saving to the total project cost.

During the later stages of the project life cycle the net saving potential will diminish as the two curves converge. In fact, some changes will not be implemented as the net effect on total cost is nil or negative.

Figure 1.1 also illustrates that any changes to the project will require alterations to documentation (drawings, specifications, bills of quantities, schedules, etc.). During the early stages of the project design, changes are unlikely to result in significant document revisions, particularly prior to the production of detailed drawings. More documentation will be produced and the potential for disruption is increased during subsequent stages of the project's life cycle.

Chapter 1 – Summary

In this chapter we have identified four main pressures which make it more difficult, and consequently even more important, to control building costs for clients.

It is equally important that if cost control is directly related to capital costs with little regard for aesthetic qualities or life cycle costs, then the buildings which result will suffer. Under these circumstances clients will not be best served or receive real value for their investment.

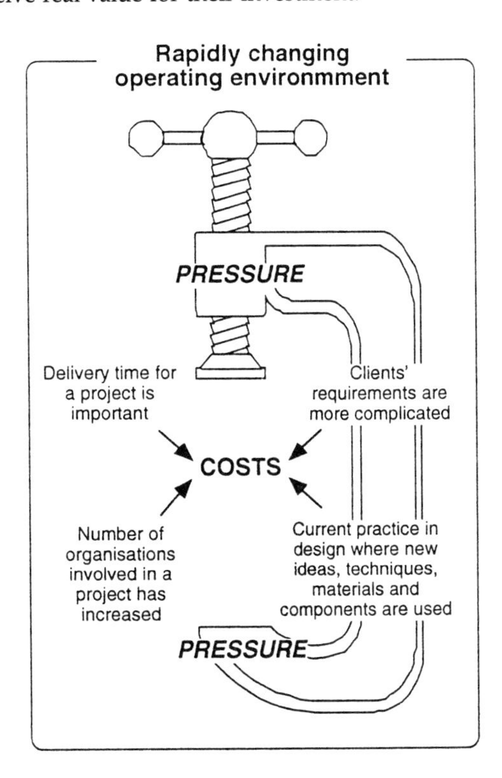

Turn to page 11

2 The Purpose of Cost Control

This chapter *is* in programmed form and you must follow the page directions exactly.

Turn to page 12

from page 11

As you work through this chapter (and the other programmed chapters) you will find it useful to have at hand a notebook and a pen or pencil.

Write down the number of each page to which you are directed, *before* you read the page. In this way, you will be able to follow your route through the programme, or find your place again easily after a break.

Also make sure you write down your answer to every question. Where a choice of answers is given, write down your choice *before* proceeding to the next page.

Feel free to make any notes from the programme as you work through it.

Turn to page 13

from page 12

In Chapter 1, you read why there was a need for cost control. This need is due to *four* main pressures:

Four pressures

(1) The delivery time for a project is important.
(2) Client's requirements are more complicated.
(3) The number of organisations involved in a project has increased.
(4) Current practice in design, where new ideas, techniques, materials and components are used.

Now that we have established the need for a cost control system, we must ask ourselves what the system should do; that is, what is the purpose of this system? Let us see how you view this problem of cost control.

Write down what you consider to be the purpose of a cost control system.

Compare your answer with the four possible answers shown below, and then turn to the appropriate page. Pick the answer which corresponds most closely to yours.

To give the client good value for money. Turn to page 20 

Value for money *Value for money does not simply mean the lowest initial capital cost. There has to be a balance between quality, fitness for purpose, initial capital cost, and the life cycle cost over the life of the element. The life cycle cost means expenditure on the running costs including maintenance, fuel, cleaning, and replacement.*

To achieve the required balance of expenditure between the various parts of the building. Turn to page 18 

Balance of expenditure *Balance of expenditure between the various parts of the building means ensuring that the cost is distributed across the elements to meet the client's and designer's requirements. For example, it may not be appropriate to spend an excessive amount on the finishings with the result that the roof specification and cost has to be minimised.*

To keep expenditure within the amount allowed by the client. Turn to page 16

All of the above three. Turn to page 14 

from page 13 You chose: **All of the above three.**

You are perfectly correct.

There are *three* purposes of having a cost control system:

(1) Giving the client good value for money.
(2) Achieving the required balance of expenditure between the various parts of the building.
(3) Keeping expenditure within the amount allowed by the client.

Important. Remember these as they will be used throughout the text.

This programme will be concerned mainly with the third purpose – keeping expenditure within the amount allowed by the client – although in achieving this third purpose, the other two purposes can also be satisfied.

This means that the final account to the client *should not* exceed the first estimate. Turn to page 19 

This means that the final account to the client *can* exceed the first estimate. Turn to page 17 

Turn to the page corresponding to the statement which you consider to be correct.

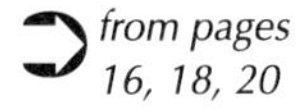
from pages 16, 18, 20

There are ***three*** purposes of having a cost control system.

This programme will be concerned mainly with the third purpose – keeping expenditure within the amount allowed by the client – although in achieving this third purpose, the other two purposes can also be satisfied.

This means that the final account to the client *should not* exceed the first estimate. *Turn to page 19*

This means that the final account to the client *can* exceed the first estimate. *Turn to page 17*

Turn to the page corresponding to the statement which you consider to be correct.

Some terminology:

Estimate	*The estimated cost based upon information that is available ranging from brief to detailed design information. A contractor estimates the cost of a project and submits a tender, which is a contractual offer.*
Forecast price prediction	*Term sometimes used in practice to represent an estimate.*
Budget	*The amount established for the project which should not be exceeded.*
First estimate	*Term we have used already which is the client's first estimate and is often used to establish the budget.*
Final account	*The cost of the project upon completion.*

from page 13

You chose:

To keep expenditure within the amount allowed by the client.

You are only partly correct, as

To give the client good value for money

and

To achieve the required balance of expenditure between the various parts of the building

are also important.

How many purposes of cost control are there?

Write down your answer.

Turn to page 15

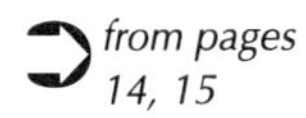
from pages 14, 15

You are incorrect.

When a client accepts a first estimate, the first estimate becomes the 'amount allowed by the client'. In a straightforward project, therefore, the final cost of the building should ***not*** exceed the first estimate.

The only possible exception to this is when the client orders additional work after the first estimate has been accepted.

The final account may of course be ***less*** than the first estimate, but this is a very rare occurrence!

Suppose a client is presented with an account for £1 000 000 when the estimate was £800 000. Has the main purpose been fulfilled?

Write down your answer.

Turn to page 21

from page 13 You chose: **To achieve the required balance of expenditure between the various parts of the building.**

Balance of expenditure *Balance of expenditure between the various parts of the building means ensuring that the cost is distributed across the elements to meet the client's and designer's requirements. For example, it may not be appropriate to spend an excessive amount on the finishings with the result that the roof specification and cost has to be minimised.*

You are only partly correct, as

To give the client good value for money

and

To keep expenditure within the amount allowed by the client

are also important.

How many purposes of cost control are there?

Write down your answer.

Turn to page 15

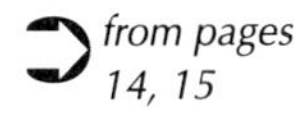
from pages 14, 15

You are correct.

When a client accepts a first estimate, the first estimate then becomes the 'amount allowed by the client', in effect the budget for the construction cost. As long as the client does not order any additional work after he has accepted the first estimate, the final account should not exceed the first estimate.

If a client is presented with an account for £1 000 000 when the estimate was £800 000, has the main purpose been fulfilled?

Write down your answer.

• ***The total cost.*** *Bear in mind that the total cost for the project will include the land cost, the design team fees, legal expenses, construction cost, taxation costs, planning and building control fees, and the cost of the interest charges on any debt capital used to finance the project. Hence, while the construction cost may be £2 000 000, the total cost of the project to the client may well be £4 500 000.* •

Turn to page 21

from page 13 You chose: **To give the client good value for money.**

Value for money	*Value for money does not simply mean the lowest initial capital cost. There has to be a balance between quality, fitness for purpose, initial capital cost, and the life cycle cost over the life of the element. The life cycle cost means expenditure on the running costs including maintenance, fuel, cleaning and replacement.*

You are only partly correct, as

To achieve the required balance of expenditure between the various parts of the building

and

To keep expenditure within the amount allowed by the client

are also important.

How many purposes of cost control are there?

Write down your answer.

Turn to page 15

from pages 17, 19

No, the main purpose has not been fulfilled.

Let us consider how the design team decides on a first estimate for a project. Basically, there are two possibilities.

One possibility is that the client can stipulate the maximum amount that can be spent on the project.

What would happen then?

The design team *would* keep their first estimate within this amount

Turn to page 22

The design team *would not* keep their first estimate within this amount.

Turn to page 24

from page 21

You said: **The design team *would* keep their first estimate within this amount.**

You are correct.

If the design team did not do this, then the client would have to reject their design as there would not be sufficient finance for building.

We have been discussing one way in which the first estimate is formulated: the client stipulates the desired cost limit and the design team examines the feasibility of keeping the first estimate within this amount.

The other possibility is that the client may give the design team a detailed description of the building he requires and asks the design team to submit an estimate.

Provided that the client accepts this estimate, will he expect the final cost to exceed this?

Write down your answer.

Turn to page 25

from page 24 **No.**

If the client did accept, they would be unable to pay the extra £80 000. In these circumstances, the client must either amend the brief or abandon the project.

We have been discussing one way in which the first estimate is formulated: the client stipulates the desired cost limit and the design team examines the feasibility of keeping the first estimate within this amount.

The other possibility is that the client may give the design team a detailed description of the required building and ask the design team to submit an estimate.

Provided that the client accepts the estimate, will he expect the final cost to exceed this?

Write down your answer.

Turn to page 25

from page 21

You said: **The design team *would not* keep their first estimate within this amount.**

You are wrong.

If the client only had £800 000 to spend, there would be no point in the design team making the estimate greater than £800 000, as this represents the maximum the client could spend on the building. The client would therefore not have accepted the estimate.

If the estimate is justifiably greater than £800 000 – that is, the building the client wants just cannot be built for £800 000 – then it is up to the design team to persuade the client either to be prepared to spend more or amend the brief to the design team.

Could the client then accept an estimate of £880 000, if the maximum funds for building are £800 000?

Write down your answer.

Experienced and inexperienced clients

It is important that the clients realise what they are getting for their investment. Clients range from organisations that have an ongoing development programme, such as a department store chain or a hospital trust, to those that build once in their lives. The first type of client is experienced and understands the industry, the latter group is inexperienced and will not fully understand what will be provided.

Turn to page 23

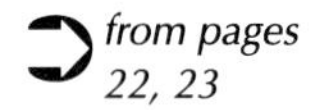
from pages 22, 23

No.

Provided the brief is not altered, he will expect the final account, on completion of the building, not to exceed the first estimate.

We have been discussing the possibilities of:

(a) the client stipulating a maximum cost, and
(b) the client giving a detailed brief to the design team to work out the cost.

Most projects, of course, lie somewhere between these two extremes.

Now, write down the purpose of cost control with which this programme will be concerned.

Turn to page 27

from page 28

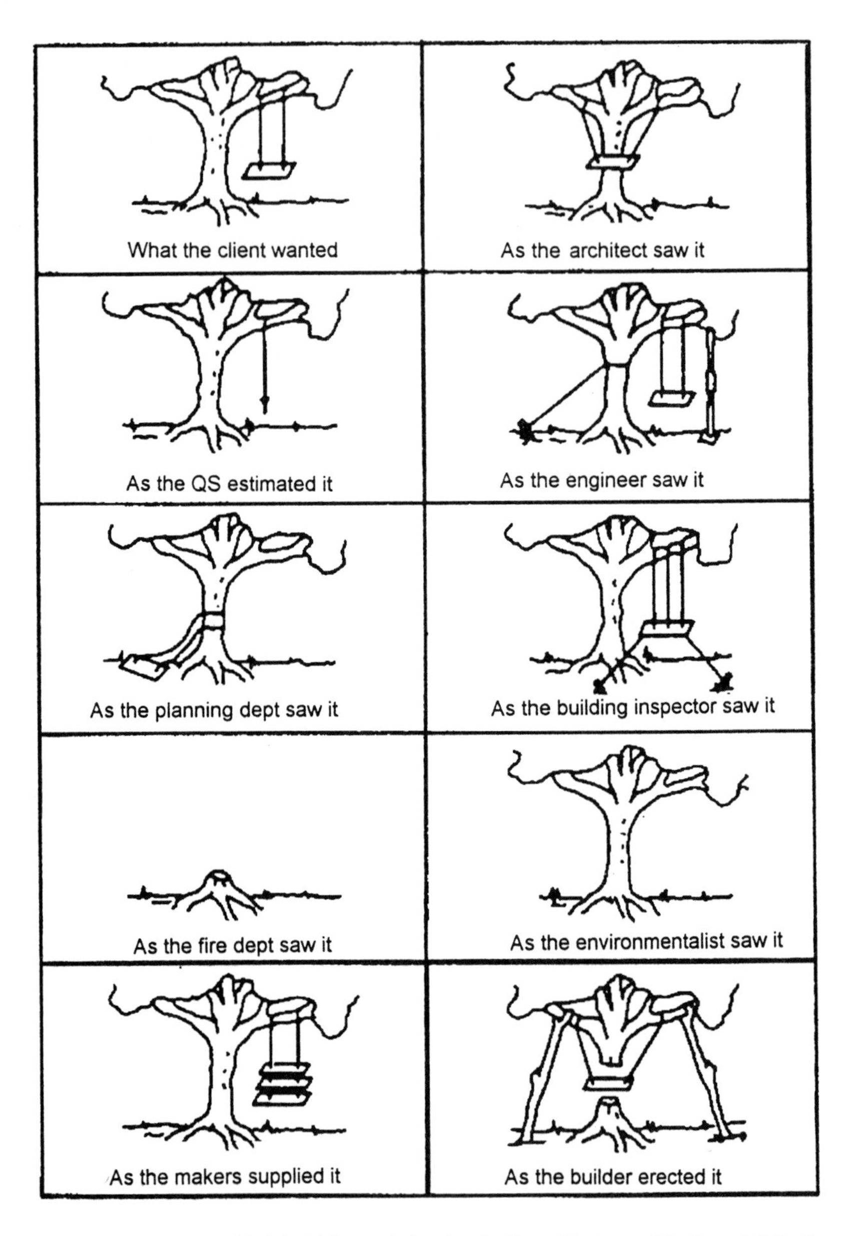

(Original idea and sketches by Dave Taylor and Anthony Walker)

Turn to page 29

from page 25 You should have written:

The purpose of cost control with which this programme is concerned is:

to ensure that the final account does not exceed the first estimate accepted by the client
and
to keep expenditure within the amount allowed by the client

This forms the objective of the cost control procedures which will be discussed in this programme.

These procedures are designed to control the effects of the four pressures mentioned in Chapter 1.

Turn to page 28

from page 27

Chapter 2 – Summary

The Three Purposes of Cost Control

(1) To give the client good value for money.
(2) To achieve the required balance of expenditure between the various parts of the building.
(3) To keep expenditure within the amount allowed by the client.

This programme is mainly concerned with the third purpose, that is, keeping expenditure within the amount allowed by the client. Thus tenders received should not exceed the amount allowed by the client. In achieving this purpose, however, the other two purposes can also be satisfied.

The ultimate aim is to ensure that the final account does not exceed the first estimate.

In this programme we only go as far as discussing cost control procedures during the design stages.

These procedures are designed to control the effects of the four pressures mentioned in Chapter 1.

Communication *Clear, concise and accurate communications are essential to ensure that all participants in the process understand what is required. The illustration on page 26 shows vividly what can go wrong.*

Turn to page 26

3 The Principles of Cost Control

This chapter *is* in programmed form and you must follow the page directions exactly.

Turn to page 30 ➲

from page 29

In the last chapter, you learned that there are three purposes of a cost control system, and that we shall be most concerned in this programme with the third purpose – that is, keeping expenditure within the amount allowed by the client.

In this programme, we shall concentrate on the design stage prior to operations on site. We will assume that the architect is responsible for leading the design team, although this is not essential as, for example, overall management could fall to an independent project manager. We are following, in principle, the RIBA *Outline Plan of Work*, and we will make reference to that document in subsequent chapters. Furthermore, we shall only deal with the processes as far as the Tender Stage. Cost control practices during construction are dealt with in other texts.

In this chapter, we will consider the nature of this system; that is, the principles of a cost control system at the design stage.

Analysis of any control system, more often than not, highlights the fact that there are three basic principles:

The Three Basic Principles

Principle One
There must be a frame of reference or set of conditions which must be adhered to.

Principle Two
There must be a method of checking or a feed-back system.

Principle Three
There must be a means of remedial action.

Important. Remember these as they will be used throughout the text.

Let us now consider these principles in the context of a cost control system.

How will a cost control system incorporate these three basic principles?

Turn to page 31

from page 30

Remember, the three basic principles of a control system are:

Principle One
There must be a frame of reference or set of conditions which must be adhered to.

Principle Two
There must be a method of checking or a feed-back system.

Principle Three
There must be a means of remedial action.

Now let us apply these principles to a cost control system in building.

Principle One (having a frame of reference) consists of *two* stages:

(1) Establishing a realistic first estimate.
(2) Planning how this estimate should be spent among the parts of the building.

Should the realistic first estimate be within the amount allowed by the client?

Write down your answer.

The first estimate *The first estimate is often the figure that clients remember most. It sets the budget. It can sometimes be based on information that is a best guess.*

Turn to page 33

from pages 35, 37

The first principle of a cost control system consists of *two* stages:

(1) Establishing a realistic first estimate.
(2) Planning how this estimate should be spent among the elements or parts of the building.

Is there a cost target for each element or work package?

Write down your answer.

A realistic first estimate

Note the term 'realistic'. Three words need to be defined: accuracy, reliability, realistic. Consider two people shooting at the targets shown in A and B below:

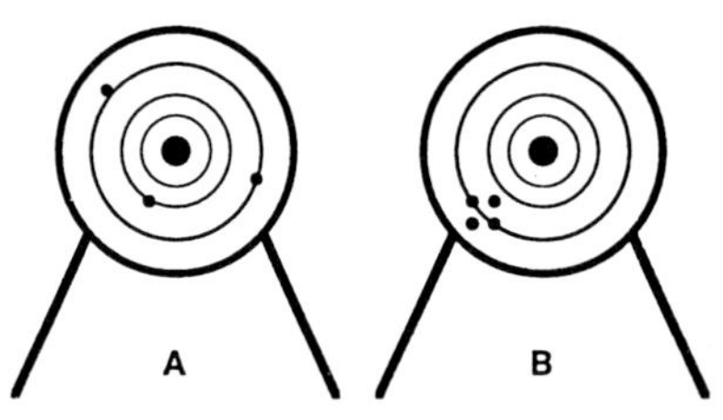

A had one shot on the bullseye and the other three are all over the target, whereas B's shots are all clustered in one place. B's shots are more considered and possibly reliable but they are not accurate as they missed the bullseye. When estimating it may be necessary to keep checking that the target is the bullseye.
A realistic estimate means that it is neither over-optimistic nor too pessimistic.

Turn to page 39

from page 31 **Yes.**

One of the purposes of a cost control system is to keep expenditure within the amount allowed by the client.

When the client has accepted the realistic first estimate, this sum is considered as the cost limit for the project. By **cost limit**, we mean the cost agreed between client and design team as the amount beyond which the project may not be pursued.

The second part of **principle one**, establishing how the estimate is to be spent among the parts of the building, entails splitting the cost limit into a number of smaller cost limits, one for each part of the building.

Will the sum of the small cost limits – usually known as 'cost targets' – equal the cost limit of the building project?

Yes. *Turn to page 35*

No. *Turn to page 37*

from page 39 **Yes.**

According to the definition, a roof is an element. Opinion differs as to what should constitute a full list of elements, but we will not indulge in this discussion here. In this book we shall use standard elements as in the BCIS (Building Cost Information Service) *Detailed Cost Analyses*.

List of elements

1 SUBSTRUCTURE
2 SUPERSTRUCTURE
2A Frame
2B Upper floors
2C Roof
2D Stairs
2E External walls
2F Windows and external doors
2G Internal walls and partitions
2H Internal doors
3 INTERNAL FINISHES
3A Wall finishes
3B Floor finishes
3C Ceiling finishes
4 FITTINGS
5 SERVICES
5A Sanitary appliances
5B Services equipment
5C Disposal installations
5D Water installations
5E Heat source
5F Space heating and air treatment
5G Ventilating systems
5H Electrical installations
5I Gas installations
5J Lift and conveyor installations
5K Protective installations
5L Communications installations
5M Special installations
5N Builder's work in connection
5O Builder's profit & attendance
6 EXTERNAL WORKS
6A Site works
6B Drainage
6C External services
6D Minor building works
7 PRELIMINARIES

Note (1) Group elements shown in upper-case letters (BCIS uses these for *concise cost analyses*).
(2) Numbering of elements follows BCIS format.

Into how many cost targets will a cost limit be split?

Write down your answer.

Turn to page 36

from page 33 You said **Yes.**

You are correct.

The sum of the cost targets ***must*** equal the cost limit.

The first principle of a control system, when applied to a cost control system, consists of two stages.

Write down the two stages.

Turn to page 32

➲ *from page 34*

In the case of the BCIS list of elements, there are 7 group elements and 30 detailed elements. You will need cost targets for the 30 detailed elements, plus targets for SUBSTRUCTURE, FITTINGS, and PRELIMINARIES, making 33 cost targets. In addition, most tender documents will often include a sum for CONTINGENCIES (for unforeseen work encountered during site operations) which should be included with the cost targets. So:

There are a possible 34 cost targets.

Managing cost by elements is used widely throughout the world. For example, the USA, Australia, Canada and South Africa all have their own list of elemental categories.

More often than not several elements will have a 'nil' value, e.g. upper floors, stairs and lift installations, in a single-storey building.

Having established a realistic first estimate or cost limit, and having planned how to distribute this estimate among the elements, we have incorporated **Principle One** of a cost control system.

We must next incorporate **Principle Two**. To do this, we must detect and measure departures from targets and thereby check that the estimate is being spent as originally established. This is called **cost checking**.

Principle Three consists of taking any necessary remedial action.
Remedial action may mean re-designing the element, changing the specificaton, or seeking an alternative solution.

When should this remedial action be taken?

Write down your answer.

Turn to page 38 ➲

from page 33 You said **No.**

You are incorrect.

If we take the overall cost limit, and split it into smaller sums of money, then adding together these smaller cost targets must give the overall cost limit.

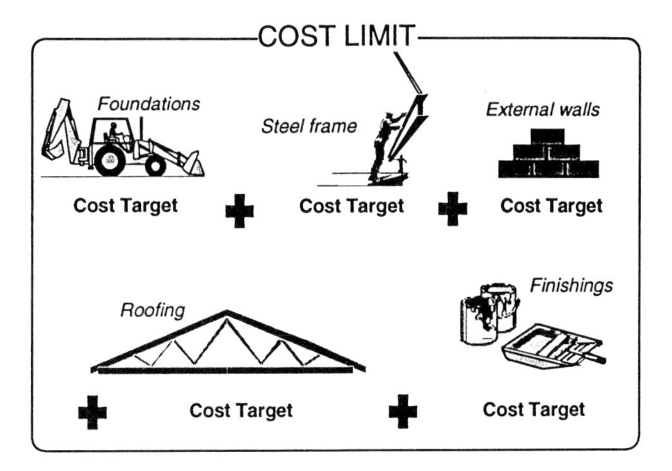

The first principle of a control system, when applied to a cost control system, consists of two stages.

Write down the two stages.

Turn to page 32

from page 36

The remedial action should be taken immediately, before the design process moves on too far.

The remedial action ensures that expenditure is contained within the amount agreed between the client and the design team, that is, the cost limit.

Occasionally, suitable remedial action is impossible, and additional finances must be obtained to complete the project. For example, the ground conditions may not be what was envisaged and a detailed site investigation may show that the foundations will need to be deeper and larger, thus incurring additional expenditure.

Who will supply the additional funds?

Write down your answer.

A note of realism *The design process is iterative, which means that the design develops as more things are discovered about the project. For example, at the early stage of design the architect may have specified a pitched roof and the outline of the roof shape. By the stage of working drawings everything will be known about the tile type, location of gutters, lap of tiles, etc. Hence, assumptions have to be made at the early stages.*

Architects and engineers do not design in elements that would not be realistic. It is the task of the cost controller to identify the elements from the drawings.

Turn to page 40

from page 32 **There *is* a cost target for each element or work package.**

Element *An element is often defined as: a major component common to most buildings which usually fulfils the same function, or functions, irrespective of its construction or specification.*

As we have mentioned previously, you do not have to use elements for the cost control of management contracts which uses work packages in order to establish how the estimate should be spent among the parts of the project. Work packages relate to the trades or skills for the contract package award, for example, the joinery works might be one work package. However, for the purposes of managing the cost at the design stage, there is still the need to focus on elements.

From here on, we will ignore work packages and only concern ourselves with **elements** as defined above.

Do you think a roof is an element?

Yes/No.

Turn to page 34

from page 38

The client will generally be asked to supply the necessary additional finance. This implies that the initial estimate was incorrect and could be an embarrassment to the design team. But the presentation to the client of a carefully completed and detailed budget is the most convincing argument available as to the necessity for additional finance.

• *Do not confuse a **budget** with **finance**. A **budget** is a financial limit. The **finance** for a project may come from a wide range of sources such as a loan from a bank, building society or insurance company, a government grant, or retained earnings.* •

Turn to page 41

from page 40

Chapter 3 – Summary

Cost control systems are based on *three* principles:

Principle One
There must be a frame of reference.

Principle Two
There must be a method of checking.

Principle Three
There must be a means of remedial action.

Having a frame of reference in a cost control system (Principle One) consists of *two* stages:

(1) Establishing a realistic first estimate.
(2) Planning how this estimate should be spent among the elements or parts of the building.

Turn to page 42

from page 41

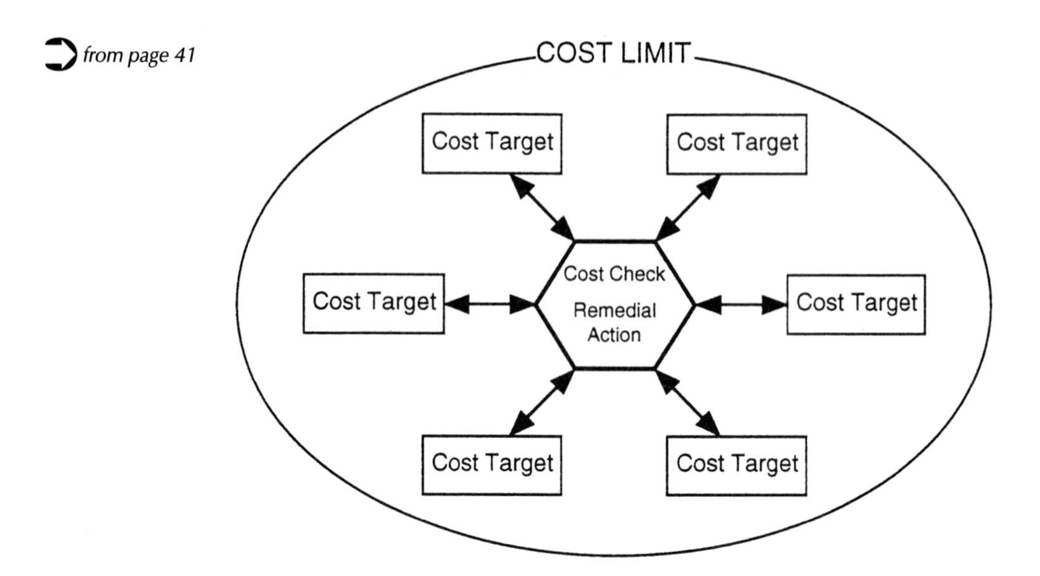

The cost limit is established and split into cost targets **Principle One**

These targets are checked as the design progresses. **Principle Two**

Remedial action is taken if necessary. **Principle Three**

Turn to page 43

4 Estimating at the Early Design Stage

This chapter ***is not*** in programmed form.

Traditional estimating techniques originated in mid-Victorian times. The techniques were designed to help the builder to calculate the tender for a project. They were not used to help a design team to prepare an estimate for a client at an early stage in building design.

During this period, general conditions such as less complex uses for buildings, lower standards of comfort and fewer amenities and little or no state control of building, led to the development of two estimating methods.

The first of these, **approximate quantities**, entailing the measurement of all the work in a proposed building, stems from a contractor being concerned solely with the price of labour, materials, plant, overheads and profit and not with the function or use of a construction project.

The second method, **single rate estimating**, such as the 'superficial floor area' method and 'cubing', evolved as clients began to demand estimates at an early stage before detailed drawings and specifications were available. The 'cube' method has been superseded by the 'superficial floor area' method (also referred to as the 'm^2' method). Single rate estimating methods, to be effective, must be used to compare costs of buildings of similar type, construction and finishes, built on similar sites under similar conditions.

Both of these estimating techniques are important and are still the methods most frequently employed today.

Providing one or two cost estimates during the design process is not sufficient. What is required is a technique of **cost control**. Neither of the two popular methods of estimating exercises any control over costs, as neither fulfils all three purposes of cost control, as stated in Chapter 2:

- ❏ to give the client good value for money;
- ❏ to achieve the required balance of expenditure between the various parts of the building;
- ❏ to keep expenditure within the amount allowed by the client;

nor incorporates all **three** principles of cost control, as stated in Chapter 3:

there must be a:

- ❏ frame of reference, establishing a realistic first estimate and planning how the estimate should be distributed among the elements;
- ❏ method of checking or feed-back;
- ❏ means of remedial action.

We will now examine traditional practices in the light of the third purpose of cost control – to keep expenditure within the amount allowed by the client – and also in the light of the principles of cost control.

Let us begin by looking at the various stages in the design process. The nomenclature we shall use in this programme is the RIBA *Outline Plan of Work*. On the next page is a condensed form of that plan together with a brief description of each stage.

Outline plan of work

Usual terminology	Stage	Purpose of work/decision to be reached	Tasks to be done
Briefing	**A Inception**	To prepare general outline of requirements and plan future action.	Set up client organisation for briefing. Consider requirements, appoint architect.
	B Feasibility	To provide the client with an appraisal and recommendation in order that he may determine the form in which the project is to proceed, ensuring that it is feasible, functionally and technically, and financially viable.	Carry out studies of user requirements, site conditions, planning, design, cost, etc., as necessary to reach decisions.
Sketch plans	**C Outline proposals**	To determine general approach to layout, design and construction in order to obtain authoritative approval of the client on the outline proposals and accompanying report.	Develop the brief further. Carry out studies on user requirements, technical problems, planning, design and costs, as necessary to reach decisions.
	D Scheme design	To complete the brief and decide on particular proposals, including planning arrangement appearance, constructional method, outline specification, and cost, and to obtain all approvals.	Final development of the brief, full design of the project by the architect, preliminary design by the engineers, preparation of cost plan and full explanatory report. Submission of proposals for all approvals.
		Brief should not be modified after this point.	
Working drawings	**E Detail design**	To obtain final decision on every matter related to design, specification, construction and cost.	Full design of every part and component of the building by collaboration of all concerned. Complete cost checking of design.
		Any further change in location, size, shape, or cost after this time will result in abortive work.	
	F Production information	To prepare production information and make final detailed decisions to carry out work.	Preparation of final production information, i.e. drawings, schedules and specifications.
	G Bills of Quantities	To prepare and complete all information and arrangements for obtaining tender.	Preparation of Bills of Quantities and tender documentation.
	H Tender action	Action as recommended in NJCC *Code of Procedure for Single Stage Selective Tendering.*	Action as recommended in NJCC *Code of Procedure for Single Stage Selective Tendering.*

It is important to be aware that there is a range of project activities spanning the life of a construction project from its inception through to completion. The RIBA Plan of Work represents a suggested set of procedures for architects during each stage of the construction project. It also embodies activities necessary for the provision of effective cost control. The outline plan of work describes the range of activities carried out up until tendering of the project, commonly referred to as the **design process**. Subsequent stages include project planning, operations on site, project completion and feed-back. The following brief description of the RIBA plan of work provides an insight into the roles of participants throughout the design process.

A **Inception**. Relates to the client's initial decision to build. During this stage the client will consider the outline requirements and appoint an architect and other members of the design team.

B **Feasibility**. During this stage the design team will attempt to establish the client's requirements in more detail. This involves the consideration of functional, technical and financial issues. Basic cost information is required in order to establish the overall viability of the project.

C **Outline proposals**. Having established the overall viability of the project, this stage determines an overall sketch design with regard to layout and construction. Alternative design forms and construction methods will be considered. The cost implications of differing design solutions will be evaluated. The outline cost plan will be prepared during this stage.

D **Scheme design**. Major issues affecting design, planning, construction method and specification will be considered at this stage. Sketch plans will cover most of the major elements of the building including structural framework, environmental engineering services and internal finishings. As the design evolves, the elements must be checked and, if necessary, earlier cost estimates revised. More accurate cost targets will be defined for all the major building elements.

E **Detail design**. During this stage all issues relating to design, specification, construction and cost are resolved. Drawings are prepared for all the building components. Cost advice will continue to be given about differing construction methods, materials and layouts. Detailed consideration should be given to the impact of maintenance and running costs associated with alternative design solutions.

F **Production information**. All final working drawings are now prepared and project schedules and specifications are finalised. Cost checks will be made to ensure that the design conforms to the cost plan.

Chapter Four – Summary

There are *two* traditional estimating procedures in common usage:

1 **Approximate quantities** Pricing composite items based on materials and labour.

2 **Single rate estimating** (e.g. superficial area method). Comparing with similar buildings.

Both are ***unsatisfactory*** unless incorporated into and used in conjunction with cost control techniques.

Problems are likely to arise where estimates are prepared without implementing a proper cost control system. Traditional estimates are only prepared and considered at the following stages:

Stage	What is done	Method used	Reason for being unsatisfactory
B **Feasibility**	First estimate prepared	Approximate quantities *or* single rate estimating (e.g. m^2 method)	First part of Principle One violated (i.e. first estimate cannot be realistic)
D **Scheme design**	Firm estimate prepared	Approximate quantities *or* single rate estimating (e.g. m^2 method)	Second part of Principle One violated (i.e. cost targets have not been planned)
H **Tender action**	Differences between tender price and first estimate explained	Pricing full Bill of Quantities	The objective of keeping the tender price within the first estimate is not achieved

In addition, as costs are not considered at the detail design stage, Principles Two and Three are violated.

The Three Basic Principles

Principle One
There must be a frame of reference or set of conditions which must be adhered to.

Principle Two
There must be a method of checking or a feed-back system.

Principle Three
There must be a means of remedial action.

Turn to page 49 ➲

5 Cost Control during Inception, Feasibility and Outline Proposals

This chapter ***is not*** in programmed form.

This chapter covers three stages of the RIBA *Outline Plan of Work*:

A Inception

B Feasibility

C Outline Proposals

In the remainder of Part 1, we will explain how the three principles of cost control are incorporated into a full cost control system. This will be appropriate for situations in which an architect is responsible for leading the client and design team. The cost control system and the information it generates follows the design process as described by the RIBA *Outline Plan of Work.* However it must be remembered that the plan of work represents an outline method of working only. It is a model and therefore will need to be modified to suit different design teams.

As stated previously, what follows will not be entirely appropriate for alternative procurement approaches, in particular, where the relationship between architect and builder and the timing of the design stage to construction phase is radically altered. However, the principles of cost control will still apply and can be adapted to suit alternative approaches.

We now consider how the three principles are incorporated into a full cost control system. The techniques as outlined will be described in more detail in Part 2 of the programme.

(As you read this chapter, you will find it useful to refer occasionally to the RIBA *Outline Plan of Work.*)

Inception Inception is the first stage of the client's initial decision to build. The design team is not yet appointed so no cost control procedures are carried out. At the inception stage the client organisation is set up for briefing. The client considers his requirements for the building and appoints a designer.

The brief *Preparing a detailed and comprehensive brief is very important. The brief states precisely what the client wants. For example, with an airport bulding, it might be the proposed number of passengers to be handled or the number of aircraft movements to be coped with. A brief can be from a few pages up to a very comprehensive document. The design team must interpret the brief to decide the type of project to be designed.*

B **Feasibility** The second stage is the feasibility stage, where the design team provides appraisals and recommendations regarding the form in which the project is to proceed.

An effective cost control system must then be set up, incorporating the three principles of cost control. The first principle to be incorporated into the system is Principle One, establishing a realistic first estimate, and deciding how to allocate this estimate among the elements.

Two approaches to setting a realistic first estimate can be considered. Firstly, the design team may be required to produce an estimate which defines the cost limit for the project. Alternatively, the design team may have to confirm that a building can be completed within the cost limit set by the client. In practice, an approach between these two extremes is usually adopted.

The information provided by the client will vary from project to project. However, whatever the approach, the function of the building is known, usually from the outset, and the location and description of the site is also available. The main factors for consideration are:

- area (the floor area of the building)
- quality (the standard of accommodation to be provided)
- shape and aesthetic features
- the constraints imposed by the planning authority on the site
- the delivery time for the project (when the client wants occupation)
- the balance between initial capital cost and the long-term costs
- cost (the likely costs).

There are other factors which can come into consideration, but for ease of demonstration, these are ignored here.

How the cost control system works

Step 1 *When a project is completed, the final account is analysed into elemental categories giving a cost per square metre for the overall project and the elemental categories. This is called a cost analysis.*

Step 2 *The cost analyses are stored in files and on computer for future reference. Remember, there is a time dimension for the analyses to be considered as demonstrated below.*

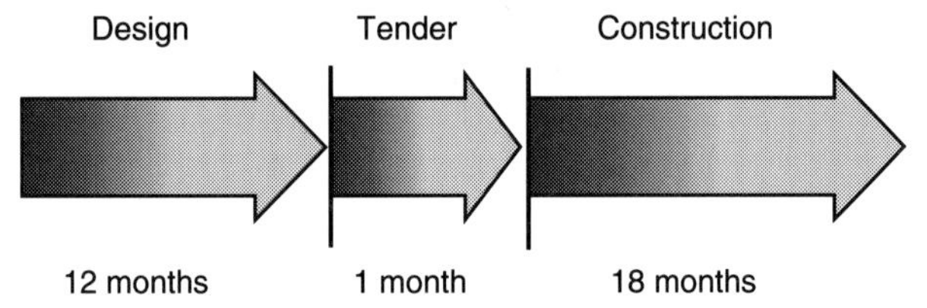

The final account is the out-turn price.

Cost planning *involves using information taken from cost analyses that have been updated.*

When either confirming or setting a realistic first estimate, consideration needs to be given to what is possible and probable. This can be done by a method of **interpolation**.

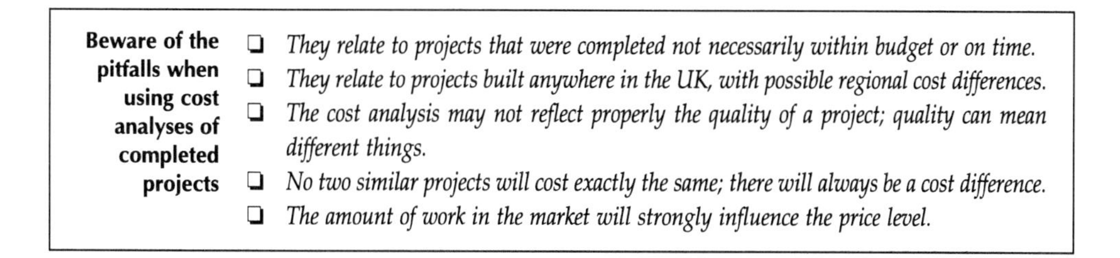

Beware of the pitfalls when using cost analyses of completed projects

- *They relate to projects that were completed not necessarily within budget or on time.*
- *They relate to projects built anywhere in the UK, with possible regional cost differences.*
- *The cost analysis may not reflect properly the quality of a project; quality can mean different things.*
- *No two similar projects will cost exactly the same; there will always be a cost difference.*
- *The amount of work in the market will strongly influence the price level.*

Interpolation method

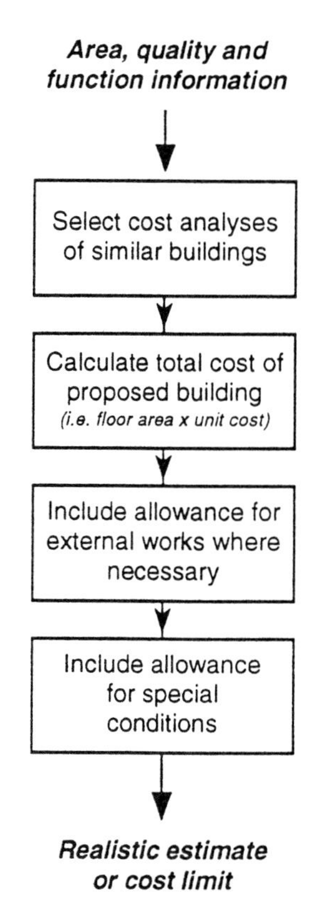

The **interpolation method** is where cost analyses of buildings of the same type are studied. This method permits the differing sizes and standards which exist between buildings to be examined and taken into account when considering costs.

The total cost of each building is generally expressed in a common unit such as cost per square metre of floor area. Thus, the total cost of the building is the cost per square metre multiplied by the total area of the floor space.

An example of part of a typical range might be:

- Cost of building A (quality X, floor space 5 000 m^2) is £800 per m^2 of floor area.
- Cost of building B (quality Y, floor space 5500 m^2) is £1000 per m^2 of floor area.

Special care needs to be taken with **external works** as published analyses may exclude costs associated with site works, drainage, external services and separate minor buildings. Consideration must also be given to any ***special conditions*** such as site problems, access, etc.

From a range of costs of similar buildings an appropriate **cost limit** can be made. In calculating the cost limit, consideration needs to be given to differences in area, quality and function and adjustments made for external works and any special conditions. This technique will be described in greater detail in Part 2 of this programme.

C **Outline proposals** The outline proposals stage is where there is still insufficient information for cost targets to be prepared for each elemental category. Nevertheless, the cost limit is established in broad terms.

Cost allocations are typically made among the major elemental groups. For example, the cost of substructures (i.e. foundations), superstructure (including internal finishes and fittings), and plumbing, mechanical and electrical services. These are calculated using a refined version of the interpolation method requiring consideration of the cost allocations of similar buildings. Such information is published, for example, by the Building Cost Information Service (BCIS), technical journals and elsewhere. Again an appropriate sum for external works and an allowance for any special conditions also needs to be included.

These allocations are adjusted to account for significant differences in the design of the buildings and for market conditions. These broad cost allocations are set out in an **outline cost plan**.

In certain situations it may be prudent to report cost information as a range of costs. This will allow the preparation of a firm estimate to be deferred until completion of the outline design. However, as we have seen, the client is usually presented with a realistic first estimate (or a confirmation of his own cost limit) at the feasibility stage.

During the outline proposal stage a variety of general design solutions will be considered. These will be modified as necessary before a general approach is agreed. Cost studies of alternative design solutions may be required and advice given on economic aspects of each solution. Although major design decisions will be taken during this stage, the project design is still far from complete.

At the end of the outline proposal stage the cost limit is confirmed. The client is given an outline cost plan with the broad cost allocations for the major parts of the building. (These matters will be explained more fully in Part 2.)

It is important to be clear about the basis of any estimate particularly when reporting figures to the client. The design team will be concerned with estimating likely tender costs, however, the client may be more concerned about his *total costs* which include, for example, professional fees, land cost and any taxes payable, such as value added tax. For this reason, make sure your report is in an appropriate format and lists all inclusions and exclusions. As far as we are concerned, our main point of reference will be the estimation of tender costs.

Chapter 5 – Summary

An appropriate system of cost control *does* incorporate the three principles of cost control. The system is set up at the **feasibility** stage, where the realistic first estimate is prepared.

At the **outline proposals** stage, broad allocations of cost are made among the major portions of the building. To do this, it may be helpful to consult various information sources, including in-house records.

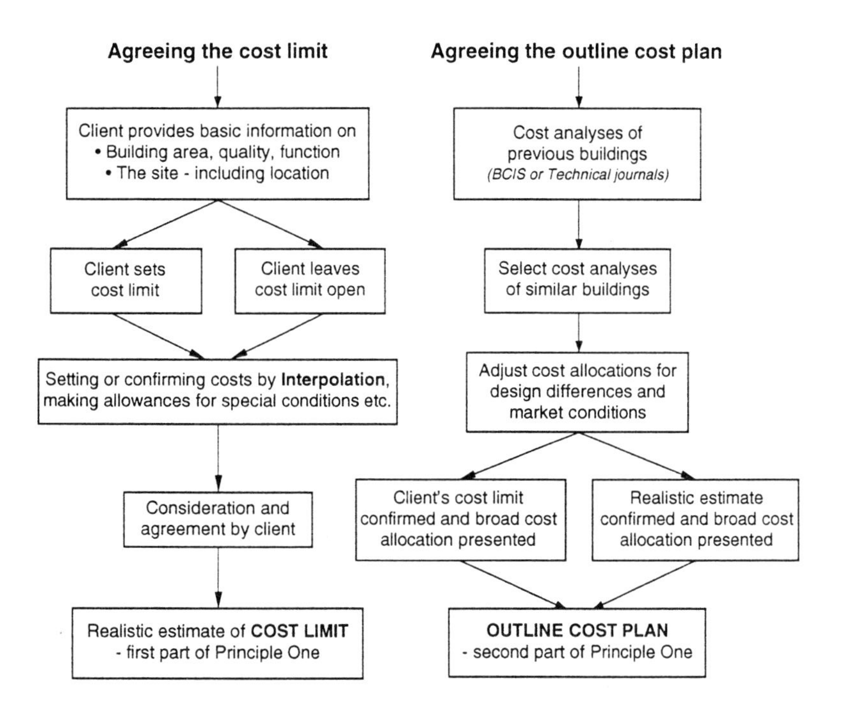

Principle One

Having a frame of reference in a cost control system consists of two stages:

- ❏ Establishing a realistic first estimate.
- ❏ Planning how this estimate should be spent among the elements of the building.

Turn to page 57 ➲

6 Cost Control during Scheme Design

This chapter ***is*** in programmed form and you must follow the page directions exactly.

This chapter covers the next stage in the RIBA *Outline Plan of Work*:

D Scheme Design

Turn to page 58

from page 57

The previous chapter described how the cost control system is set up at the **feasibility** stage and is developed at the **outline proposals** stage.

What is the next design stage?

Write down your answer.

Turn to page 60

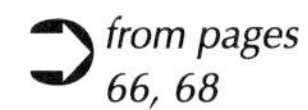
from pages 66, 68

You said **Yes.**

You are incorrect.

The cost plan and the traditional estimate have little in common, save that they are sometimes prepared in the same way. They differ on three major points.

Let us look at these differences in detail:

1 A traditional estimate is a prediction of the cost of a specific design. The cost plan is a statement of how the design team proposes to distribute the available money among the elements of the building. This allows firm decisions regarding design detail and specification to be postponed until a later stage (i.e. during the detail design stage).

Is the cost plan specific for a particular design?

Write down your answer.

Turn to page 64

from page 58 The next stage is the **Scheme design** stage.

The brief is completed and the design developed. It should take account of the requirements of the planners, the appearance, construction method, outline specification and cost. At this stage all necessary approvals are obtained, such as approvals from local planning authorities, and approval that the drawings and specifications are in compliance with the Building Regulations and fire codes.

During the scheme design stage, sketch plans are produced and the **cost plan** is formulated. (This is sometimes referred to as the **detailed cost plan** – we will continue to refer to it as the cost plan.) During this stage comparative cost studies may be required.

The **cost plan** is a statement of how the design team proposes to distribute the available money among the elements of the building.

Notice the distinction between the outline cost plan prepared during outline proposals and the cost plan produced during scheme design:

- The **outline cost plan** is a statement of how the design team proposes to distribute the available money among the ***major*** portions of the building (i.e. substructure, superstructure, internal finishes, fittings, plumbing, mechanical and electrical services, external works, preliminaries, and contingencies).
- The **cost plan** comprises all the appropriate elements (i.e. drawn from the 34 cost targets as outlined in Chapter 3).

An example of an outline cost plan is given on page 253, and a cost plan is given on page 281.

In preparing the cost plan a cost target is set for each element.

What should the sum of the cost targets equal?

Write down your answer.

Turn to page 62

from pages 66, 68

You said **No.**

Correct.

Although the cost plan and the traditional estimate are sometimes prepared in the same way, they do not have the same function. They differ on three major points.

Let us look at these differences in detail.

1 The first major difference is that a traditional estimate is a prediction of the cost of a specific design.

The cost plan is a statement of how the design team proposes to distribute the available money among the elements of the building. This allows firm decisions regarding design detail and specification to be postponed until a later stage (i.e. during the detail design stage).

Is the cost plan specific for a particular design?

Write down your answer.

Turn to page 64

from page 60

We might expect the sum of the cost targets to equal the cost limit. However, at this stage, it is advisable to make an allowance for fluctuations in price level and unexpected changes in design between scheme design and receipt of tender. This is known as price and design risk. This allowance is very important as unexpected rises in wages, materials or components, together with possible design difficulties, sometimes make it impossible to keep the final cost of each element within its cost target.

Therefore, in general, the sum of the cost targets plus the price and design risk allowance will equal the cost limit.

During the scheme design stage sketch plans are produced concurrently with the detailed cost plan. These sketch plans will be more detailed than in earlier stages, showing the locations of all walls, windows, doors, partitions, staircases, wash basins, etc.

Should these sketches be produced with the cost limit and outline cost plan in mind?

No. Turn to page 68

Yes. Turn to page 66

from page 64

The two principles are:

Principle One
There must be a frame of reference (set up at the Scheme design stage).

Principle Two
There must be a method of checking (set up at the Detail design stage).

3 The third major difference is that the traditional estimate in no way helps the design team to detail the design within a cost framework.

Cost plan allowances are independent of a specific design. The cost plan remains valid until a satisfactory design solution is obtained or until the design team decides to amend the allowance.

In addition, the cost plan is valid for any detailed design provided the design team does not alter the chosen standards of quality and is prepared to modify any details which exceed the cost targets.

With a cost plan the design team (can/can not) change the allotment of money between elements within the overall cost limit.

Write down your answer.

Turn to page 65

from pages 59, 61

No. A cost plan *is not* specific for a particular design.

As we have said, there are *three* major differences between the cost plan and the traditional estimate.

2 The second major difference is that once the estimate is prepared, it is seldom referred to until the time comes to compare it with the lowest bid for the work.

The cost plan is used continuously throughout the **Detail design** stage as a method of checking that the detail design is contained within the cost limit. The cost plan is therefore a frame of reference. This frame of reference is set up at the **Scheme design** stage.

Write down the two principles of a cost control process which are incorporated when a cost plan is used.

Turn to page 63

from page 63

The design team **can** change the allotment of money between elements within the overall cost limit. (This will be explained fully in Chapter 7.)

The cost plan differs from the traditional estimate in ***three*** major ways:

(1) The cost plan is not specific for a particular design.
(2) The cost plan is referred to continuously throughout the design process.
(3) The cost plan helps the design team to detail the design within a structured cost framework.

Do you think the production of a cost plan is a joint enterprise by the whole design team?

Yes/No.

Write down your answer.

Turn to page 67

from page 62 **Yes.**

You are correct.

The outline cost plan must be used as it states how the design team proposes to distribute the available money on the major elements of the building.

In preparing the cost plan, we must bear in mind that only total areas and numbers are known, together with an indication of quality. At this stage the design team is not prepared to enter into detailed consideration of the specification. Items such as the types of doors and windows will not be specified in detail at the scheme design stage. We must acknowledge this situation in preparing the cost plan and understand what is required of the cost plan.

The cost plan is often prepared using similar techniques to those used in preparing the traditional estimate.

Will the cost plan serve the same function as the traditional estimate?

Yes. *Turn to page 59*

No. *Turn to page 61*

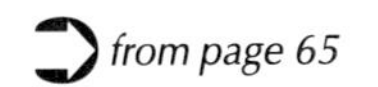
from page 65

Yes.

All members of the design team must participate in the formulation of the cost plan and have to be committed to it. Every time a line is put on a drawing there is a cost incurred. While some members of the design team may feel the cost limit for their particular design responsibility is too low, they must accept the limit or argue the case with the team and the client on the need to raise the limit or take a sum from one of the other elements.

Remember, the cost limit is made up of a number of cost targets – one target per element.

The nature of the construction project may require specialist consultants to give advice. For instance, in the design of a concert hall there is the need for acoustics expertise to advise on sound.

Specialist consultants (should/should not) co-operate in the formulation of the cost plan.

Write down your answer.

Turn to page 71

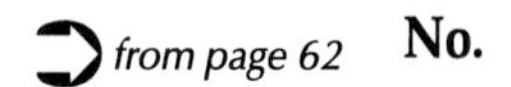
from page 62

No.

You are incorrect.

Sketch plans must be produced with the cost limit and outline cost plan in mind. The outline cost plan is a statement of how the design team proposes to distribute the available money on the major portions of the building.

In preparing the cost plan we must bear in mind that, only total superficial areas and the general layout are known, together with an indication of quality. At this stage the design team will not be in a position to enter into detailed consideration of the specification. Items such as the types of doors and windows will not be specified in detail at the scheme design stage. We must acknowledge this situation in preparing the cost plan and understand what is required of the cost plan.

The cost plan is often prepared using similar techniques to those used in preparing the traditional estimate.

Will the cost plan serve the same function as the traditional estimate?

Yes. *Turn to page 59*

No. *Turn to page 61*

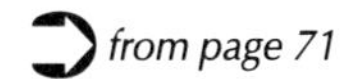
from page 71

Yes.

When consultants have been properly involved in setting targets, the design team can justifiably expect construction costs to fall within these targets.

In the USA, if the lowest tender is not within a certain percentage (usually 10%) of the estimate given to the client, then the design team must modify the design to meet the budget – they will not be reimbursed for the time spent on the exercise. It would not be unreasonable for a UK client to expect the same service.

During the scheme design stage the design team finalises the scheme design and produces statements of quality standards and functional requirements for each element which are translated into a cost target.

Is this process of translation carried out under traditional estimating practices?

Yes/No.

Write down your answer.

Turn to page 70

from page 69

No.

This process represents increased responsibility to ensure the cost targets for each element are realistic and reflect the specification and quality required.

Turn to page 72

from page 67

Specialist consultants **should** co-operate in the formulation of the cost plan.

It would be unreasonable to expect specialist consultants to accept a cost target within their area of responsibility unless they had consented to that target during the formulation of the cost plan.

Should the design team expect attainment of cost targets concerning specialist consultants?

Yes/No.

Write down your answer.

Turn to page 69

from page 70

Chapter 6 – Summary

At the **scheme design** stage, the **cost plan** (i.e. detailed cost plan) is formulated, often in a similar way to the traditional estimate. Thereby, Principle One (having a frame of reference) is incorporated. Sketch plans are prepared with the cost plan in mind.

The **cost plan** is simply a statement of how the design team proposes to distribute the available money on the elements of the building.

The cost plan differs from the traditional estimate in *three* ways:

(1) The cost plan is not specific for a particular design.
(2) The cost plan is referred to continuously throughout the design process.
(3) The cost plan helps the design team to detail the design within a structured cost framework.

The cost plan is, in addition, a joint enterprise by the whole design team including the specialist consultants.

The design team's decisions on each element are translated into cost targets, whose sum, together with the design and price risk allowance, equals the cost limit.

Turn to page 73

7 Cost Control during Detail Design

This chapter *is* in programmed form and you must follow the page directions exactly.

This chapter covers the fifth stage of the RIBA *Outline Plan of Work*:

E Detail Design

Turn to page 75

from page 77

The design team must be informed of the cost for each element.

The estimated cost of each fully designed element is compared with the corresponding cost target expressed in the cost plan.

This is a cost check for each element.

What principle of the cost control system has thus been incorporated?

Write down your answer.

Turn to page 76

from page 73

Up until now, we have been *planning* what should be done with the money available for a project. During *detail design* we take action; at this stage final decisions must be made on every matter related to design, specification, construction and cost. Every part and component of the building must be fully designed and all designs must be completely **cost checked**. Remedial action must be taken if necessary.

What cost control principle must therefore be incorporated into the process at this stage?

Write down your answer.

Turn to page 77

from page 74

Principle Two (there must be a method of checking).

If cost checking is not done, then all previous work is wasted and useless.

What principle of the cost control system has still to be incorporated?

Write down the principle.

Turn to page 78

from page 75

Principle Two (there must be a method of checking) and Principle Three (there must be a means of remedial action) must be incorporated.

Detailed designs are prepared for each element of the building. In many countries, including the UK, the design team must also ensure that the design complies with the construction safety legislation. All designs must be complete, materials and components selected and construction problems must be solved before work may safely proceed to the next stage of design – *Production information.*

Production information means that there is sufficient detail for the contractor and specialist contractors to build the project. The contractor needs reinforcement schedules, door and ironmongery schedules, and window schedules. The specialist contractors will often prepare their shop drawings from the design information. For example, a joinery shop needs details about how every piece of timber will be cut and assembled.

In reality, a project cannot be 100% designed prior to work starting on site because information is missing. Computer aided design (CAD) systems have helped, but they can only provide information that is input.

However, the drawings prepared by the design team do not have to be polished working drawings for use on the site. Dimensioned sketches and notes in whatever form are sufficient, provided all design solutions are complete.

Since detailed designs and specifications are available, the most accurate and appropriate estimating technique is the approximate quantities method. The approximate quantities are measured for each element and priced. Prices for some specialist work may be provided by specialist contractors. For instance, the lift manufacturer may provide the price for the lift installation with due allowance being made for the builder's work.

These estimates are then checked against the corresponding cost targets in the cost plan.

Who should then be informed of the cost for each element?

Write down your answer.

Turn to page 74

from page 76

Principle Three (there must be a means of remedial action).

This remedial action must take place within the time and budget allowance so that the objective of keeping expenditure within the amount allowed by the client is achieved.

Once again if this is not done, all previous work is wasted.

Let us consider the situation where cost checking reveals that remedial action is necessary. Two courses of remedial action are possible.

1. If the cost of the element design is greater than a *realistic* cost target, the element design should be changed so that it is within the cost target.

If the anticipated cost of the electrical installations is £120 000, and the cost target was £100 000, what would be the recommendation to the design team?

Write down your answer.

Turn to page 80

from page 80

No.

The overall cost limit should ***not*** be altered. This is the whole point of cost planning – keeping expenditure within the amount allowed by the client. The cost plan can be changed at this stage as long as the cost limit is not altered.

Of course, if the cost of the element design is just within the cost target then the cost limit will not be exceeded.

Do you think any action should be taken in this case?

Yes/No.

Turn to page 81

from page 78

It would probably be recommended that the element design should be changed so that it is within the cost target.

For example, wiring and lighting layouts could be re-examined for economies.

Let us now look at the second course of remedial action.

2 If the cost check of the element design shows that the cost target is *unrealistic*, cost targets should be adjusted throughout the cost plan, thereby releasing surplus funds from other elements.

In doing this will the overall cost limit be altered?

Yes/No.

Turn to page 79

from page 79 **Yes.**

The design team must be informed, preferably in writing.

If the cost of the element design is just within the cost target, the design is confirmed in writing as being suitable for the preparation of production drawings.

If the cost is significantly within target, then surplus funds may be released for other elements.

Each element is treated in turn this way and any necessary action is taken and completed. Certain elements are more cost significant or are more likely to give trouble than others, so the cost checks on these elements are completed first. This will be explained more fully in Part 2.

If remedial action produced a building of low quality, would the design team be justified in continuing without consulting the client?

Yes/No.

Turn to page 83

from page 83

The cost limit may be overspent, therefore the main purpose of cost control may not be achieved.

The final cost check is therefore an important part of the cost control procedure at the **detail design** stage.

If this final check confirms that all is well, the project will proceed to production information, bills of quantities, tender. Not all projects will use a bill of quantities, but they will all require a schedule of materials, labour and plant.

These processes will *not* be dealt with in this programme.

Turn to page 84

from page 81

No.

In cases where remedial action would produce a building of low quality, the design team should ask the clients if they wish to release additional funds.

In this ***undesirable situation***, the cost plan is a powerful tool with which to argue the case for more funds.

Once all element designs have been cost checked, a final overall cost check is carried out. The aim must be to ensure nothing has been overlooked.

What may happen if a final cost check is not performed and, as a result, one or two of the elements have been overlooked?

Write down your answer.

Turn to page 82

from page 82

Chapter 7 – Summary

At the detail design stage, **Principle Two** (there must be a method of checking) and **Principle Three** (there must be a means of remedial action) are incorporated into the cost control system.

Flow of design team work at **detail design** stage:

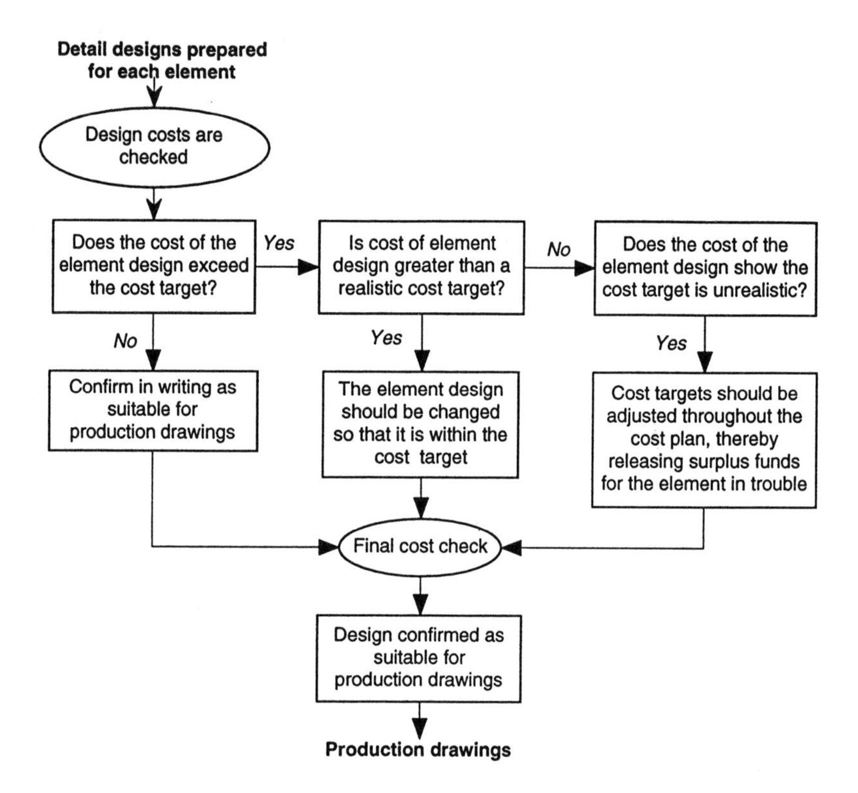

You have now completed Part 1 of this programme.

To revise this part of the programme before going on to Part 2, answer the questions on page 85.

You can find out how well you have learned the material in Part 1 by comparing your answers with the answers given on pages 86 and 87.

Part 1 Test

1 Why is the control of building costs just as important now but more difficult than in the 1980s?

2 How many purposes has a cost control system?

3 State the fundamental purpose of a cost control system through which the other purposes can be achieved.

4 What are the three basic principles of a cost control system?

5 Define in your own words:

- ❑ cost limit
- ❑ cost check
- ❑ cost target
- ❑ element
- ❑ cost plan.

6 At which stage in the design process is:

(a) a cost control system set up?
(b) the cost plan prepared?

7 What basic information about a project does a client usually give to the design team when asking them to prepare a first estimate?

8 In how many ways does the cost plan differ from the traditional estimate? (List the differences.)

9 Write down the cost control principle (or principles) incorporated at the **detail design** stage.

10 Draw a rough diagram to show the flow of design team work at the **detail design** stage.

Answers overleaf

Answers to Part 1 Test

		Marks	
1	Client's requirements are more complicated.	(1)	
	Number of organisations involved in a project has increased.	(1)	
	Delivery time for a project is important.	(1)	
	Current practice in design where new ideas, techniques, materials and components are used.	(2)	
			5
2	Three.	(1)	
			1
3	Keeping expenditure within the amount allowed by the client *or*	(3)	
	Ensuring the final account does not exceed the first estimate.		**3**
4	**Principle One**: Having a frame of reference *or* establishing a realistic first estimate and planning how the estimate should be spent among the elements.	(3)	
	Principle Two: Having a method of checking.	(2)	
	Principle Three: Having a means of remedial action.	(2)	
			7
5	The **cost limit** is the cost of a project to the client, beyond which the project may not be pursued.	(2)	
	A **cost check** is performed by comparing the approximate quantities estimate of a completely designed element with its cost target in the cost plan.	(2)	
	A **cost target** is the allowance made for an individual element at the scheme design stage.	(2)	
	An **element** of a building is a major component common to most buildings which usually fulfils the same function or functions irrespective of its construction or specification.	(3)	
	The **cost plan** is a statement of how the design team wishes to spend the available money on the elements of the building.	(3)	
			12
6	(a) Feasibility.	(1)	
	(b) Scheme design.	(1)	
	Note: An outline cost plan is prepared during **outline proposals**.		**2**

Turn to page 87 ➲

from page 86

		Marks	
7	Area, quality, function (any order).	(3)	
			3
8	(1) The cost plan is not specific for a particular design.	(3)	
	(2) The cost plan is referred to continuously throughout the design process.	(2)	
	(3) The cost plan helps the design team to detail the design within a structured cost framework.	(2)	
	In addition the cost is a joint enterprise by the whole design team including the specialist consultant members.	(1)	
			8
9	**Principle Two** (there must be a method of checking).	(1)	
	Principle Three (there must be a means of remedial action).	(1)	
			2

10

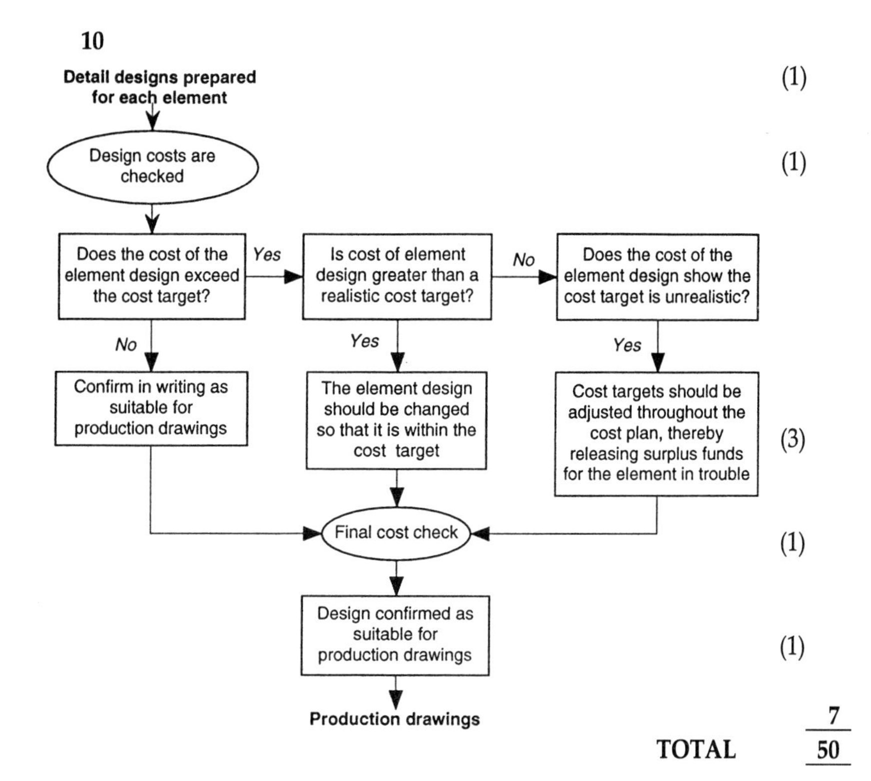

7

TOTAL **50**

Part 2

The Techniques of Cost Control

Introduction

In Part 1, you learned the purpose and the principles of cost control. In Part 2, you will learn some of the techniques which are used to control building costs during the design process.

We must make it clear that the costs discussed throughout this part of the book are initial *capital* costs and that we make no attempt to consider full life cycle costs (the running costs including maintenance, repair, cleaning, energy, insurances, etc.). In practice, capital, maintenance, and operating costs must all be considered in any comprehensive system of economic appraisal. However, such considerations are outside the scope of this programme.*

There is another facet of the cost–design interaction which is outside the scope of this programme. Building is usually undertaken for a financial or social return. It is often necessary to measure the benefits which the building or facility will give. In complex cases, this may be done by preparing a cost benefit analysis. This technique considers both tangible and intangible costs and benefits. For example, a cost benefit study was carried out for the Jubilee Line extension to Docklands to take account of the benefits to individuals and to London's economy as a whole.

No matter what economic appraisal techniques are employed, it will be necessary to prepare estimates of initial capital cost and to control the development of a design within the agreed cost estimate. Capital cost planning can, therefore, be used either on its own or supplemented by the more refined techniques outlined above.

Turn to page 91 ➲

* For further reading on this subject, see *Life Cycle Costing for Construction* and/or *Life Cycle Costing: Theory and Practice*, both by R. Flanagan and G. Norman.

8 Essential Constituents of Elemental Cost Analysis

This chapter *is* in programmed form and you must follow the page directions exactly.

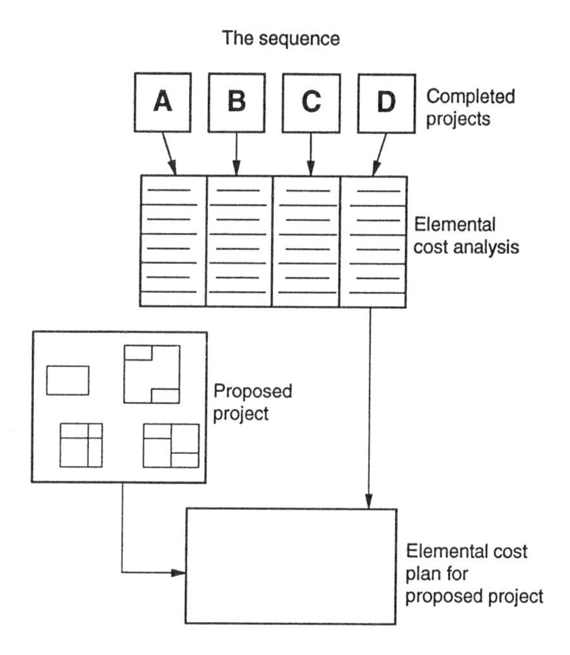

Turn to page 92

from page 91

An elemental cost analysis provides a meaningful yardstick for clients and the design team about the cost of similar projects. For example, by examining the cost of the roof, external walls, windows and external doors we can get an indication of how much it costs to enclose a building and keep out the weather.

Examining cost analyses

Examination of only one cost analysis yields useful information. However, because of the variations between buildings, it is obviously essential to be able to compare the cost analysis of several different buildings of the same type. This allows a wider appreciation of the distribution of costs.

What is needed is a uniform method of analysis and measurement for all analysis. The method of analysis described here is that used by the Building Cost Information Service (BCIS) of the Royal Institution of Chartered Surveyors (RICS).

Principles, instructions and definitions of this method are set out in the *Standard Form of Cost Analysis (SFCA)*, published by the RICS. Originally, three levels of information were published, namely, *concise*, *detailed* and *amplified* forms of cost analysis. Currently, only two levels, *concise* and *detailed* cost analyses, exist. This leaves a void for users of these cost analyses and many organisations seek to rectify this deficiency by collecting and keeping more detailed in-house records. See, for example, the amplified details recorded under 'specification and design notes' as shown in Appendix A.

Before discussing the essential constituents of an elemental cost analysis, let us summarise the general purpose of this form of analysis and briefly discuss a few of its specialised uses.

Turn to page 93

from page 92

The purpose of any cost analysis is to reveal the distribution of the costs of a building among its elements ***in terms which are meaningful to both designers and clients***. This allows the costs of two or more buildings to be compared.

As you can see from this statement, elemental cost analysis can obviously be used in several ways. We now discuss four of the simplest uses, under the headings of:

- appreciation
- judgement
- belated remedial action
- planning.

Appreciation

As mentioned before, one cost analysis can be used to give clients and designers an appreciation of how the costs of the elements compare with one another and with the total cost of the building.

Judgement

This appreciation can then be used in judging which elements have disproportionate cost when considered in relation to the efficiency and quality which they contribute to a building. Hence ideas as to how costs could have been allocated to obtain a more balanced design can be developed.

How can a cost analysis help the design team when a tender is found to be greater than the first estimate?

Write down your answer.

Turn to page 95

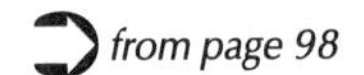
from page 98

Cost

Cost
Elements
Yardstick

The building costs analysed in an elemental cost analysis could be either the costs to the contractor or the costs to the client (i.e. the contractor's 'prices'). As the basic purpose and use of elemental cost analysis are all for the benefit of the design professions and the client rather than the contractors, it is obviously the client's costs that are analysed.

The actual costs which a client has paid for a completed building can be obtained from the final account prepared on project completion – an accountant often refers to this as the out-turn cost. The priced bill of quantities submitted as a tender by a contractor will usually contain a number of provisional or prime cost sums which may not be exactly the same as the final cost of the items they represent.

Do you think the costs analysed should be those on the tender or those on the final account?

Cost and price *There is often a debate about cost and price. The contractor incurs costs in constructing the project to which are added overheads and profit; the client is then charged a price for the work. We accept this definition, but for the purpose of clarity in this book, the term* **cost** *has been used throughout.*

There is nothing scientific about costs. Any organisation will use the standard approach of ***labour*** *+* ***materials*** *+* ***plant and equipment*** *+* ***overheads*** *+* ***profit*** *to arrive at a charge. The profit allowance reflects a number of factors, such as what the market will bear at that point in time.*

Tender. Turn to page 99

Final account. Turn to page 102

from page 93

If a tender is found to be greater than the first estimate, a cost analysis of the tender would show which of the elements gave rise to disproportionate costs. A rather belated substitute for cost planning could then be performed. These elements could then be re-designed in an attempt to bring the cost down. This use can be summarised as follows.

Belated remedial action

Cost analysis can be used to reveal the sources of over-expenditure during 'inquests' on receipt of tender, and thus allow belated remedial action to be taken.

What obvious use of cost analysis (that is, obvious in the context of this programme) has not yet been mentioned?

Write down your answer.

Turn to page 100

from pages 99, 102

(a) The costs analysed are the client's costs on the tender.

Elements

Cost Elements Yardstick

Elemental cost analysis seeks to relate costs to those functions of buildings which are common to all buildings (i.e. functions like 'keeping out the weather' rather than 'teaching children' or 'making beer barrels').

Why is a 'trade-by-trade' or a 'common arrangement by work sections' analysis, as can be used in priced bills of quantities, unsuitable for relating costs to function?

Write down your answer.

Turn to page 103

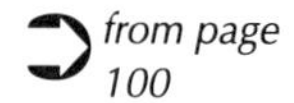
from page 100

Any analysis is performed to increase our understanding of how a *whole* is made up of constituent parts.

When deciding on a method for analysing a whole project, the essential constituents need to be determined. In our case these would be:

(a) which characteristic is to be analysed,
(b) the sub-units which are to be considered,
(c) the most useful way of expressing the results.

The purpose of elemental cost analysis determines that **cost** is the characteristic to be analysed. **Elements** (i.e. functional components) are the sub-units of buildings which should be considered. Since part of the purpose is to allow comparisons of the costs of two or more buildings, the best method of expressing the result will be in terms of a common **yardstick**. This allows comparison of the costs of buildings of different sizes to be made without making repeated reference to the sizes of the buildings.

Write down three words which summarise the essential constituents of elemental cost analysis.

Turn to page 98

from page 97

The essential constituents of elemental cost analysis are:

- ❑ **Cost** the characteristic to be analysed,
- ❑ **Elements** the sub-units used,
- ❑ **Yardstick** the most useful way of expressing the results.

Cost
Elements
Yardstick

The remainder of this chapter is devoted to a discussion of each of these essential constituents.

Turn to page 94

from page 94 You chose: **tender.**

You are correct.

But the argument presented on page 94 would lead the uninitiated to judge that the costs analysed should be those on the final account.

If you would like to read the argument which we have given for the analysing costs on tenders instead of those on the final account, turn to page 102 now.

If you are already convinced that it is better to analyse costs on tenders, choose from the following list the most accurate description of the costs analysed in elemental cost analyses:

(a) The costs analysed are the client's costs on the tender.
(b) The costs analysed are the contractor's costs on the final account.
(c) The costs analysed are the client's costs on the final account.

Write down your answer.

Turn to page 96

from page 95

Planning

Cost analyses may be used as cost information during the design of future building projects.

Some of these uses are so closely related that you may well argue that they should not be considered separately: we beg to differ! The first two are useful to clients and designers during the synthesis of their appreciation of how costs are related to function and design – both 'passive' uses. The other two are useful in helping to control costs – 'active' uses.

Thus the general purpose of elemental cost analysis and the four uses mentioned in this chapter can be summarised as follows:

Purpose *The purpose of elemental cost analysis is to reveal the distribution of the cost of a project between its elements in terms which are useful to both designers and clients and, by so doing, to allow the costs of two or more buildings to be compared in a meaningful way.*

Uses *Cost analysis can be used to:*

*(a) enable clients and designers to appreciate how cost is distributed among the functional components of a building (**appreciation**);*

*(b) enable clients and designers to develop ideas as to how costs could have been allocated to obtain a more balanced design (**judgement**);*

*(c) allow remedial action to be taken on receipt of high tenders, by revealing the sources of over-expenditure (**belated remedial action**);*

*(d) help with the cost planning of future building projects (**planning**).*

Turn to page 97

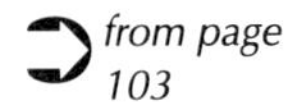
from page 103

You chose **(b)**.

You are correct.

An 'element' may be thought of as simply a *building* functional component such as roof, windows and external doors, external walls, etc. (The term element does not describe a *user's* functional unit such as a classroom, a gymnasium, a hotel bedroom, a hospital ward, etc.)

In Part 1, an element was defined as

'a major component common to most buildings which usually fulfils the same function, irrespective of its design, specification or construction.'

This definition is worth remembering for practical purposes although not all of the elements in current use are accurately described by this definition.

The elements we shall use in this programme are those used by the Building Cost Information Service (BCIS) in their *Detailed Cost Analyses*.

Study the detailed cost analysis given in Appendix A and write down any of those elements which you consider are *not* accurately described by the definition given above.

Turn to page 104

from page 94 You chose: **final account.**

You are incorrect.

From the preceding argument, it would certainly *appear* that the costs on the final account should be analysed in preference to those on the accepted tender.

In practice, the costs on priced bills of quantities submitted on tender are analysed. The reason for this apparently irrational choice is that costs on final accounts are not significantly different from those on tenders. Therefore, costs on tenders are analysed because they are available much earlier. This, of course, has the disadvantages that the provisional or prime cost sums on the tender may not be exactly the same as the final cost of the items. Nevertheless, the advantage of tender costs being up-to-date is generally considered to outweigh these disadvantages.

Choose from the following list the most accurate description of the costs analysed in elemental cost analyses:

(a) The costs analysed are the client's costs on the tender.
(b) The costs analysed are the contractor's costs on the final account.
(c) The costs analysed are the client's costs on the final account.

Write down your answer.

Turn to page 96

from page 96

Analysis of the costs incurred by each trade is ***unsuitable for relating costs to function*** because many elements include work executed by several trades. Additionally, any one trade may work on a number of elements. For example, a bricklayer may work on external walls, internal partitions, and external works. Similarly, common arrangement by work sections do not relate to functions (i.e. elements).

Because of this, the purpose of elemental cost analysis cannot always be achieved by using priced bills of quantities in their original form. A separate analysis has to be produced by gathering together the costs of each element, which are usually spread over several pages and under separate headings in a bill of quantities.

Furthermore, a bill of quantities may not necessarily have been used. Converting from the bill into elements can be a major exercise. Some organisations have produced elemental bills to overcome this, but they are unpopular with contractors because they price in trades or work packages.

Which of the following is a list of building elements?

(a) Classrooms, gymnasium, a hotel bedroom Turn to page 105

(b) Roof, windows and external doors, external walls Turn to page 101

from pages 101, 105

'Preliminaries' and 'contingencies' are certainly not components of completed buildings. Note, however, that they could be regarded as fulfilling the same functions during the design and construction of different buildings!

A useful characteristic of elements is that they tend to be self-explanatory to all concerned, including clients. However, when preparing or using an analysis, items included in each element must be clearly identified. Unfortunately, there is no room in this programme to include a full specification of the items which should be included in each element.

The problems of delineation, however, are relatively minor and you will be able to understand everything in this programme without having a full specification available.

Write down our definition of a building 'element'.

Turn to page 106

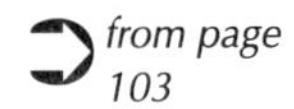
from page 103

You chose **(a).**

You are incorrect.

Classrooms, gymnasiums, and bedrooms are all *user's* functional units rather than *building* functional components (i.e. elements). Roof, windows and external doors, external walls and stairs are all considered as elements by most authorities.

In Part 1, an **element** was defined as:

'a major component common to most buildings which usually fulfils the same function, or functions, irrespective of its design, specification or construction'.

This definition is worth remembering for practical purposes, although not *all* of the elements in current use are accurately described by this definition.

The elements we shall use in this programme are those used by the Building Cost Information Service (BCIS) in their *Detailed Cost Analyses.*

Study the detailed cost analysis given in Appendix A and write down any of those elements which you consider are *not* accurately described by the definition given above.

Turn to page 104

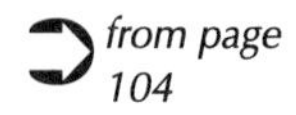
from page 104

A building 'elemental category' or 'element' may be defined as 'a major component common to most buildings which fulfils the same function, or functions, irrespective of its design, specification or construction'.

Thus, for example, 'roof' is an element regardless of the fact that we may be considering a tiled pitched roof or an asphalt flat roof.

Yardstick

Cost
Elements
Yardstick

It follows from the general purpose of elemental cost analysis that we must have a method of expressing costs which will allow sensible comparisons between buildings *of different sizes and uses*. It is obviously futile to try to compare the costs of buildings directly without first expressing them in a way which takes account of the differences in size or capacity.

Costs of any one type of school can usefully be considered in relation to the number of pupil-places provided and, 'cost per place' is a useful yardstick to use when comparing the costs of schools of different sizes. 'Cost per place' is an example of what BCIS calls a 'functional unit cost'. Note that the word 'functional' in this context refers to the use to which the whole building is put. Other examples of functional unit cost are 'cost per m^2 of usable floor area' for office buildings, 'cost per bed' for hospitals, 'cost per seat' for theatres, etc.

Expressing costs as 'functional unit cost' is, however, *not* used as the standard method of expressing the results of elemental cost analyses.

Write down one good reason for this.

Turn to page 108

from page 113

(b) £40.50 per m^2 is incorrect.

£40.50 per m^2 is the cost of the upper floors per m^2 of *upper floors*. You were asked to calculate the cost of the upper floors per m^2 of 'floor area', which in this context always means *gross floor area*.

The example is repeated below. When you have calculated the correct answer, turn to page 115.

Example

In a building of gross floor area 1700 m^2, the upper floors have an area of 800 m^2 and cost £32 400. Calculate the cost of the upper floors per m^2 of floor area, correct to two decimal places.

Turn to page 115

from page 106

'Functional unit costs' are not used as the standard method of expressing elemental cost analysis because they would not allow comparisons of the costs of buildings which have different uses (e.g. schools, sports centres, office blocks).

Furthermore, even when comparing similar buildings such as schools, functional units tend to cloud important issues concerning quality, quantity and other factors.

A perfect cost yardstick would allow meaningful comparisons to be made between buildings of similar type, design, quality, date of tender, etc. and, at the same time, also allow comparisons with buildings of different use classifications.

The costs of buildings will obviously increase as the volume enclosed by the building increases and gross floor area increases. So both 'cost per cubic metre' and 'cost per metre^2' are worth considering for use as the standard yardstick.

Are clients (and designers) more likely to think of buildings in terms of volume or floor area?

(a) Volume. *Turn to page 109* 

(b) Floor area. *Turn to page 111* 

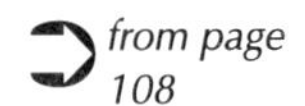
from page 108

You chose **(a)**.

You are incorrect.

As long as each room is of usable height, the height of each storey is not generally as important to clients as the gross floor area. Clients are therefore more likely to think in terms of floor area rather than in terms of volume.

In addition, costs are not affected as much by an increase in the height of each storey as they are by an increase in superficial area. In other words, costs have a closer relationship to floor area than to volume. Because of this, design teams also think in terms of floor area.

'Cost per m^2 of gross floor area' (often expressed as '£ per m^2 gfa') is, therefore, the most realistic yardstick to use for comparing costs. The costs in costs analyses are therefore expressed as 'cost per m^2 of gross floor area'.

By referring to Appendix A, write down the *units* used for our yardstick of cost.

Turn to page 113

from page 115

$$\text{Cost of upper floors per m}^2\text{ of floor area} = \frac{\text{cost of upper floors (£)}}{\text{gross floor area (m}^2\text{)}}$$

$$= \frac{\text{£32 400}}{1700\,\text{m}^2}$$

$$= \text{£19.06 per m}^2$$

(to two decimal places)

Note the distinction between this and 'cost of upper floors per m^2 of *upper* floors', which would be found by evaluating:

$$\frac{\text{cost of upper floors (£)}}{\text{area of upper floors (m}^2\text{)}}$$

Return to page 115

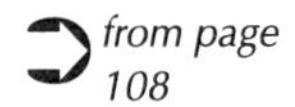
from page 108

You chose **(b)**.

You are correct.

Clients and designers are more likely to think of buildings in terms of floor area than in terms of volume. As long as each room is of usable height, the height of each storey is not generally as important to clients as the floor area.

In addition, costs are nearer to being proportional to floor area than to volume, because costs are not affected as much by an increase in the height of each storey as they are by an increase in floor area.

'Costs per m^2 of gross floor area' ('£ per m^2 gfa') is, therefore, the most realistic yardstick to use for comparing costs. This is the method used for expressing the results of elemental cost analyses. You may also see, elsewhere, this expressed as 'costs per m^2 of gross *internal* floor area'.

By referring to Appendix A, write down the *units* used for our yardstick of cost.

Turn to page 113

from page 113

(a) The total cost of the building is expressed as cost per m^2 of gross floor area by calculating

$$\frac{\text{total cost of building (£)}}{\text{gross floor area of building (m}^2\text{)}}$$

In Appendix A,

$$\text{total cost per m}^2 \text{ of gross floor area} = \frac{£1\,153\,793}{1408\,\text{m}^2}$$

$$= £819.46 \text{ per m}^2 \text{ of gross floor area}$$

(b) The cost of each element per m^2 of gross floor area is found by calculating

$$\frac{\text{cost of element (£)}}{\text{gross floor area of building (m}^2\text{)}}$$

In Appendix A,

$$\text{cost of external walls per m}^2 \text{ gfa} = \frac{£27\,069}{1408\,\text{m}^2}$$

$$= £19.23 \text{ per m}^2 \text{ of gross floor area}$$

Return to page 113

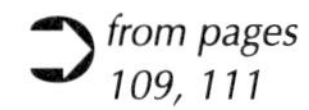
from pages 109, 111

All the costs per m² of gross floor area are expressed in £, correct to two decimal places.

If you would like to see how the costs in column 4 in Appendix A are calculated, turn to page 112.

Note that the cost of each element is divided by the *gross* floor area of the building, (*and not just the floor area appropriate to the element*). This is especially easy to forget when dealing with elements such as '2B: Upper floors'.

The 'gross floor area' is always the *total* area of *enclosed* floor space. No deductions are made for partitions, stairwells, corridors, cupboards, etc. Measurements are taken to the internal face of the enclosing walls (i.e. external walls). The definition given by the *Standard Form of Cost Analysis* is:

Gross floor area

(1) Total of all enclosed spaces fulfilling the functional requirements of the building measured to the internal structural face of the enclosing walls.
(2) Includes area occupied by partitions, columns, chimney breasts, internal structural or party walls, stairwells, lift wells, and the like.
(3) Includes lift, plant, tank rooms and the like above main roof slab.
(4) Sloping surfaces such as staircases, galleries, tiered terraces and the like should be measured flat on plan.

Note:
(a) Excludes any spaces fulfilling the functional requirements of the building which are not enclosed spaces e.g. open ground floors, open covered ways and the like. These should be shown separately.
(b) Excludes private balconies and private verandas which should be shown separately.

Example
In a building with a gross floor area of 1700 m², the upper floors have an area of 800 m² and cost £32 400. Calculate the cost of the upper floors per m² of floor area, correct to two decimal places.

Which answer is correct?

(a) £19.06 per m² Turn to page 115

(b) £40.50 per m² Turn to page 107

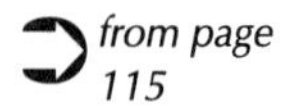
from page 115

Some clients who deal with buildings of identical function, find 'functional unit cost' is a useful method of comparison. Hence, the cost of the actual building per functional unit is given in cost analyses for such purposes.

Type of building	Functional unit cost
School	Cost per pupil place
Hotel	Cost per bedroom
Hospital ward	Cost per bed space
Office	Cost per m^2 usable floor area. (Usable floor area is that area fulfilling the main functional requirement of the building and excludes for example all circulation areas. In a commercial context it usually equates to that area used for the calculation of rent.)

Revision

As revision for this chapter, we would like you to think carefully about the distinctions between the *purpose* of elemental cost analysis (i.e. the general reason for such an analysis being performed), the nature of the analysis (i.e. how you would describe it to someone else), and the specialised *uses* to which analyses may be put.

(a) Compose and write down the definition of elemental cost analysis which you would use to explain it to someone who had never come across the term.

(b) Write down the purpose of elemental cost analysis.

Turn to page 116

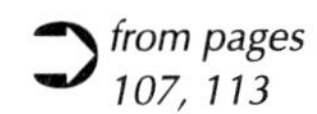

(a) £19.06 per m^2 is the correct answer.

(If you would like to check your working, turn to page 110.)

'Cost per m^2 of gross floor area' is not a *perfect* cost yardstick. The costs of buildings are not *exactly* proportional to floor area – even for buildings which differ only in size. Costs per m^2 of floor area should be used with discretion. We know there will be differences in shape, height, ground conditions, external works and a host of other things, but we still need something as a basis for considering projects. As we will discuss in the next chapter, variations in quantity, quality and price level cause differences in cost per m^2 of gross floor area.

Nevertheless, cost per m^2 of gross floor area relate costs to client's principal requirement and also go a long way towards eliminating variations in cost due to differences in the size of buildings.

Throughout the world everybody uses the cost per m^2 as the method of communicating cost.

The following extract is from a typical cost analysis:

Functional unit	**Rate**
1325 m^2 usable floor area	£827.08

Why do you think this 'functional unit cost' has been included?

Write down your answer.

Turn to page 114

from page 114

Chapter 8 – Summary

(a) **Elemental cost analysis is the analysis of the costs to clients which are given on tenders, to determine the probable cost of each element of a building.**

(b) **The purpose of elemental cost analysis is to show the distribution of the costs of a building among its elements in meaningful terms to both clients and design teams and, by so doing, to allow the cost of two or more buildings to be compared.**

It is unlikely that your choice of words will be exactly as the above: this does not matter, as long as the meaning is the same.

(c) **Write down the uses of elemental cost analysis mentioned in this chapter.**

Turn to page 117

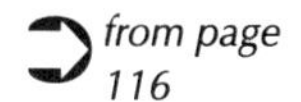
from page 116

(c) The uses of elemental cost analysis as discussed in this chapter are:

Appreciation
Elemental cost analysis can be used to enable clients and designers to appreciate how cost is distributed among the elements of a building.

Judgement
Elemental cost analysis can be used to enable clients and designers to develop ideas as to how costs could have been allocated to obtain a more balanced distribution of cost for the design.

Belated remedial action
Elemental cost analysis can be used to allow remedial action to be taken on receipt of tenders above the budget, by revealing the sources of over-expenditure.

Planning
Elemental cost analysis can be used to help with the cost planning of future construction projects.

(d) List the three essential constituents of elemental cost analysis.

(e) Why are elements employed in the analysis of building costs?

(f) Why is a yardstick used in elemental cost analysis?

(g) What yardstick of cost is usually employed?

Write down your answers.

Turn to page 118

from page 117

(d) The three essential constituents of elemental cost analysis are:

- ❑ **cost**
- ❑ **elements**
- ❑ **yardstick.**

(e) Elements are used as the sub-units of analysis so that costs can be related to those functions of buildings which are common to all types of buildings.

(f) A yardstick is used in elemental cost analysis so that costs of buildings of different sizes can be compared without explicitly considering the sizes of the buildings.

(g) 'Cost per m^2 of gross floor area' is used as the yardstick of cost.

Turn to page 119

9 Factors Affecting Element Costs

This chapter *is* in programmed form and you must follow the page directions exactly.

Turn to page 120 ➲

from page 119

So far we have only mentioned one of the factors which affects the costs of elements: the size of the building. Expressing the cost of each element as 'cost per m^2 of gross floor area' allows naive, but nevertheless useful, comparisons of the costs of buildings of different size. However, such comparisons, can only *isolate* the differences in costs per m^2. They cannot *account* for these differences.

For a complete understanding of costs, all of the factors affecting element costs should obviously be studied so that adjustments can be made to account for the effect of each of the factors.

Before we consider in detail how this is done, see if you can postulate one or more possible reasons for the difference between the element costs in the following example.

Example

In the building analysed in Appendix A, internal doors cost £5.87 per m^2 of gross floor area (element 2H).

If analysis of another building gave a figure of £9.34 per m^2, what possible reasons for this difference can you think of?

Write down your answers.

Turn to page 122

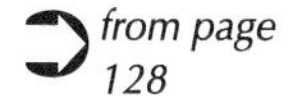
from page 128

(a) Wherever possible, analyses include element unit quantities in order to provide a measure of the amounts of each element in different buildings. When comparing buildings, we can judge how much of the variations in element costs is due to differences in the amounts of each element.

(b) Since element unit rates are relatively unaffected by variations in Quantity, they isolate the differences in cost due to Quality and Price level.

If, however, there is a very large increase in the Quantity of an element, the element unit rate would probably decrease because cheaper rates are usually obtained for large quantities. Economies associated with large quantities affect all resources. For example, labour rates can become more competitive with the use of more efficient gang sizes, continuity of work, repetition and learning curves, etc. Savings in material costs are possible with discounts for bulk buying, use of full loads rather than half loads, etc. Mechanical plant can be used to advantage and methods of working can also show savings in element unit rates.

Therefore, when using the analyses in cost control, selected projects should be reasonably compatible both with regard to type and size. These points are considered again in later chapters.

Turn to page 131

from page 120

There are several possible reasons, the most probable are:

(a) There may be a larger number (*quantity*) of internal doors per m^2 of gross floor area in the second building.

(b) The doors of the second building may be of better *quality* than those in the first.

(c) The tender for the second building may have been prepared a few years after the tender for the first, and the increase in cost may be at least partly due to an increase in the general market *price level.*

It is, of course, highly likely that the increase is due to a combination of these three major factors and some others.

A large amount of time and effort could obviously be spent trying to analyse each of the factors affecting element costs. In practice however, it is found that most of the difference between the costs of any one element in different buildings can usually be accounted for by considering just three major factors:

Factor 1 Quantity.
Factor 2 Quality.
Factor 3 Price level.

Important. Remember these as they will be used throughout the text.

When the actual adjustments are made, *price level* is normally taken first, followed by *quantity*, and then *quality*. However, this order has been changed on the following pages in order to make the explanations clearer.

To determine the effect of these factors on the cost of each element, an elemental cost analysis must obviously contain a specification of the quantity and quality of each element together with the date of tender and other information relating to price level.

Turn to page 127

from page 126

Example

The total cost of the internal walls and partitions for building A is £80 000 and for building B is £480 000. Let us assume that all of the difference between these costs for the two buildings is due to Quantity.

Then, comparison of the costs of this element per m^2 of gross floor area will eliminate the difference in costs due to the differing sizes of the two buildings, but it will not eliminate the differences in costs due to the difference in density of that element in the two buildings.

	A	*B*	Ratio *A/B*
Total cost of 'internal walls and partitions'	£80 000	£480 000	1/6
Gross floor area (m^2)	1000	3000	1/3
Cost per m^2 of gross floor area	£80	£160	1/2

In other words, using this comparison eliminates the difference in costs due to the size of the buildings; and thus any remaining difference in cost must be due to a difference in density of the element (remember we are assuming Quality and Price level do not affect cost). Further investigation will confirm this.

	A	*B*	Ratio *A/B*
Area of 'internal walls and partitions'	2000 m^2	12 000 m^2	1/6
Density*	2	4	1/2

$$^*\text{Density} = \frac{\text{area of internal walls and partitions}}{\text{gross floor area}}$$

The ratio of total costs is the same as the ratio of areas of internal walls and partitions because all of the difference between the element costs is due to Quantity. Eliminating the 1 to 3 difference in floor areas by expressing the costs as 'costs per m^2 of gross floor area' brings the ratio down to 1 to 2. In this case, all of this remaining difference should be due to the variation in the 'densities' of internal walls and partitions: this is confirmed by the ratio of densities, which also turns out to be 1 to 2.

In most cases, however, the ratio of 'costs per m^2 of gross floor area' would not be the same as the ratio of the 'densities' because the costs per m^2 would also reflect variations in Quality and Price level.

What way of expressing costs will isolate the difference in cost due to the combined effect of Quality and Price level?

Write down your answer.

Turn to page 130
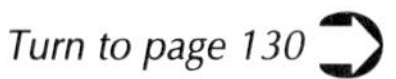

from page 127

In Appendix A, the Quantity of wall finishes is 904 m^2.

In practice the measurements currently used as element unit quantities (as recommended in the *Standard Form of Cost Analysis*) are satisfactory for most elements. Thus, comparison of the Quantity of an element in two buildings gives a reliable indication of how costs are affected by this factor.

When comparing the costs of two or more buildings, valuable knowledge will remain hidden if only the total Quantities of each element are compared, for example, the knowledge of how much of a change in element unit quantity is due to a difference in the size of the buildings and how much is due to a difference in the density of the element in the buildings remains undisclosed.

For instance, it may be tempting to say, 'The element – "internal walls and partitions" – is more expensive because there are more internal walls and partitions'. This does not really tell us very much: there may be more internal walls and partitions because one building is larger than the other, or because one contains a higher proportion of small rooms, or both.

It is therefore useful to analyse Quantity into two factors: the Quantity due to the building size, and the Quantity due to density of the element. The first of these can obviously be assessed by comparing the gross floor areas.

How can the relative 'densities' of internal walls and partitions in two buildings be assessed?

Write down your answer.

Turn to page 126

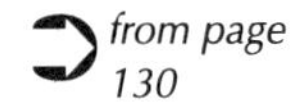
from page 130

Element	Element unit rate
2B Upper floors	**£32.14 per m^2**
2H Internal doors	**£131.27 per m^2**
5D Water installations	**£480.73 per draw-off point**

When analyses are being used, it is obviously vital that confusion never occurs between 'cost of element per m^2 of gross floor area' and 'element unit rate'. Confusion is most likely with elements such as internal doors, when the latter becomes 'cost of internal doors per m^2 of internal doors' and the former is 'cost of internal doors per m^2 of gross floor area'. Both are expressed in £ per m^2.

To help to avoid this confusion, we shall always refer to

$$\frac{\text{cost of element}}{\text{gross floor area}}$$ as **cost of element per m^2 of gross floor area**

and

$$\frac{\text{cost of element}}{\text{quantity of element}}$$ as **element unit rate**.

Important. Remember these as they will be used throughout the text.

Example

Write down clearly labelled values, for both of the measures of cost mentioned above, for upper floors (element 2B) in Appendix A.

Turn to page 128

from page 124

The relative 'densities' of internal walls and partitions can be assessed by comparing the values of the ratio

$$\frac{\textbf{Quantity of internal walls and partitions}}{\textbf{gross floor area}}$$

in the two buildings.

Detailed analyses include details of 'specification and design notes'. However, there is a tendency for insufficient information to be given and as a consequence these ratios are rarely included. Nevertheless, there is no reason why individual organisations should not collect and record this information for their own in-house cost analyses.

These ratios, as well as being useful when considering how Quantity has affected the costs of an element in two or more buildings, can also be used during the design of a new project. For instance, comparison of the window-to-floor ratios for alternative design solutions *for a given floor area* will help to give a quick appreciation of the relative economy of the alternative design solutions.

As we have already discussed on page 124, it is often useful to consider how the total effect of Quantity can be attributed to the difference between the sizes of the buildings and the difference between the densities of the element in the buildings. This can be done either by comparing the gross floor areas and the densities, as we have seen, or by eliminating the effects of each factor in turn.

Expressing costs as 'costs per m^2 of gross floor area' eliminates (for our purposes) the difference in costs due to the size of the buildings, but it is still misleading, for we are ignoring the effects of different 'densities'. If Quality and Price level do not affect costs, then, comparison of the costs of the element per m^2 of gross floor area will *not* eliminate the difference in costs due to the difference in density of that element in the two buildings.

Turn to page 123

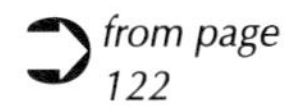
from page 122

We will now discuss each of the major factors in turn.

Factor 1 Quantity

The 'quantity' of an element is the 'amount' of the element in the building. It is determined by the client's requirements of floor area, span, loading, etc., and by the architect's design solutions to these requirements.

As the total cost of an element obviously depends on the quantity of the element, some measure of the quantity should be given in an analysis so that the variations in costs due to changes in quantity can be assessed.

For many elements, finding a suitable unit of measurement to reflect quantity within the building presents few problems. Thus, for different buildings which have identical quality and price levels but varying quantities of an element, straightforward comparisons and useful conclusions might be made. However, in practice this ideal 'measure of quantity' has not yet been found for all elements. It is, perhaps, too much to hope that even with further study, a number, area, or ratio will be found which will provide a simple index of the relative size of each element. This is particularly evident with environmental engineering services elements.

The term **Element Unit Quantities** (often abbreviated to EUQ) is commonly used to describe these simple measures of quantity. In places within the text we have simply referred to **Quantities** (with a capital Q), rather than using the term 'Element Unit Quantities'.

Write down the Element Unit Quantity of wall finishes given in Appendix A, (element 3A).

Turn to page 124

from page 125

Cost of upper floors per m^2 of gross floor area = £16.07
Element unit rate for upper floors = £32.14 per m^2

Note The first of these is £16.07 per m^2 of *gross* floor area and the second is £32.14 per m^2 of ***upper floors only***.

Element unit rates will be used during the preparation of the cost plan in later chapters.

Why are

(a) element unit quantities, and
(b) element unit rates given in cost analyses?

Write down your answer.

Turn to page 121

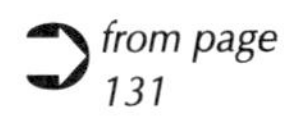
from page 131

There is one aspect of 'quantity' which we have not yet considered.

Even when the element unit quantities for a given element are the same, a difference in the quantities of the items *within* the element may cause a difference in the cost of an element. For instance, 90 m^2 of glazed doors and 10 m^2 of flush doors would probably be more expensive than 10 m^2 of glazed doors and 90 m^2 of flush doors. However, both examples when totalled have an element unit quantity of 100 m^2.

In cost analyses, where is information given about the quantities of various items within each element? (See Appendix A.)

Write down your answer.

Turn to page 132

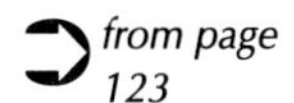
from page 123

Expressing costs as a ratio of cost of element to quantity of element will highlight the difference in cost due to the combined effect of Quality and Price level. This ratio is called the 'element unit rate'.

This method takes account of the quantity of an element as shown below:

$$\frac{\text{cost per m}^2\text{ of gross floor area}}{\text{density}}$$

Expressed in full gives:

$$\frac{\text{cost of element}}{\text{gross floor area}} \times \frac{\text{gross floor area}}{\text{quantity of element}}$$

which, when simplified, gives:

$$\frac{\text{cost of element}}{\text{quantity of element}}$$

This ratio (for our purposes) should have an approximately constant value for any one element if none of the *other factors* affecting costs varies from building to building *except* the Quantity of the element. (Remember we have decided that the major factors under consideration are: Quantity, Quality and Price level.)

In Appendix A, the 'element unit rates' are given in column 6.

Example

As you can see in Appendix A, the Quantity of '2C: Roof' is measured in m^2, so the element unit rate is £71.66 per m^2.

By referring to Appendix A, write down the element unit rates, including units, for:

- ❑ **2B Upper floors**
- ❑ **2H Internal doors**
- ❑ **5D Water installations.**

Turn to page 125

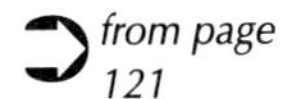
from page 121

Summary of Quantity

The cost–element unit quantities relationship can be considered either in two stages or in one stage, as shown below.

(a)

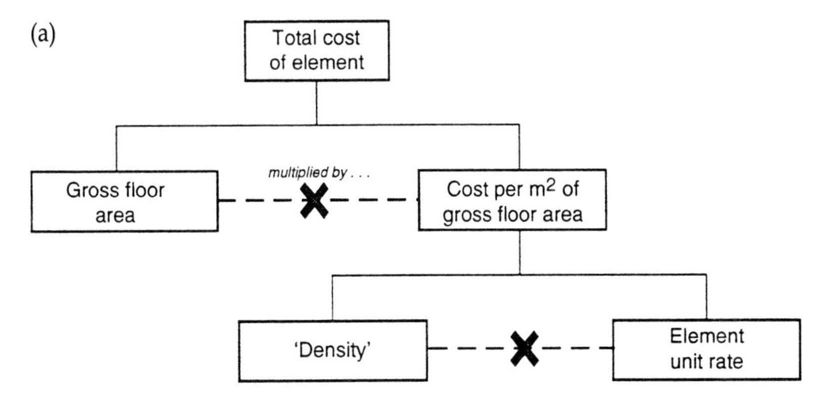

where density is element unit quantity divided by gross floor area, and cost per m^2 of gross floor area relates to the cost of the particular element being considered.

(b)

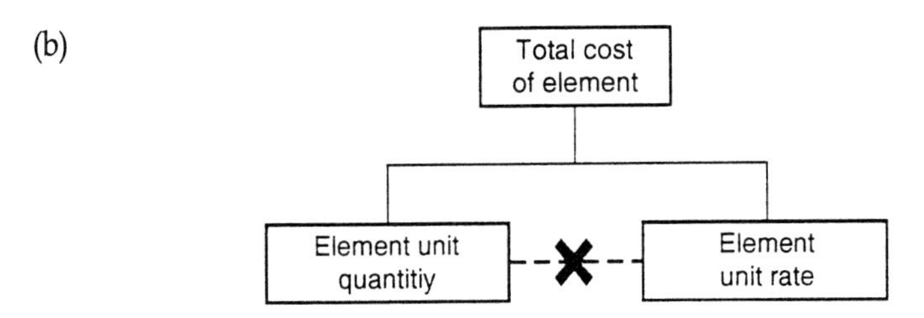

Comparison of element unit quantities gives a measure of how much of the variation in element costs between buildings is due to differences in the amount of the element.

Element unit rates isolate the differences in cost due to Quality and Price level. For these reasons use of element unit quantities and element unit rates gives a more refined answer.

Turn to page 129

from page 129

With a detailed analysis, the only place to include such information is within the 'Specification and design notes'.

Where various forms of construction or many different finishes are contained within one element, further breakdown should be given. It is suggested that the net areas and costs of the key types of construction (or finishes) should be given separately in the 'Specification and design notes' section. Indeed, it would be prudent to record the total costs, if these are significant, which must include all the costs of items pertaining to that construction.

In practice, differences in cost due to differences in the quantities of items within an element are often considered as being due to '**Quality**'.

We shall now go on to consider **Quality** and **Price level**.

Turn to page 133

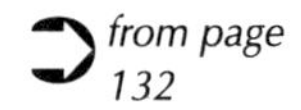
from page 132

Factor 2 Quality

Changes in Quality obviously affect the costs of elements. However, the relationship between Quality and cost is far more complex than Quantity and cost.

What does 'Quality' mean? There is quality of design, quality of materials, quality of workmanship, quality of performance, quality of service provided to the client. Hence the idea of quality has several facets.

Perhaps easiest to understand is the quality reflected in choice of materials, for example, the type of facing bricks, the choice of finishes, etc. This is very much a question of specification. However, exact details can be hidden where simple element unit quantities do not disclose the make-up of items within the element. Much of the foregoing might relate to aesthetic qualities. Performance requirements must also be considered. As examples, consider the wearing characteristics of floor finishes, or, U-values for roofs, or then again it might be the level of engineering services required by the client. Again these design solutions would normally be reflected in specification notes.

Quality of workmanship is even harder to establish. Part of it would be specified within the detailed contract documentation (not available with a cost analysis), and part of it would be implied in the expectations of certain clients or for certain building types.

Some organisations have tried to rank 'quality' on a scale of 1 to 10. However, no simple or universal quality index has, as yet, been devised for any element. The extent of changes in cost due to variations in quality can be isolated by an experienced cost consultant from cost analyses or bills of quantities of the buildings under consideration. For example, adjusting for the variation in cost between softwood and aluminium windows can be done from knowledge of market prices. Making allowances for different levels of engineering service requirements would be more difficult and often more cost significant. In practice, because of the make-up and complexity of design, much reliance is placed on professional judgement.

What part of a detailed analysis would have to be inspected to isolate the information necessary to assess how Quality had affected costs?

Write down your answer.

Turn to page 138

from page 137

You chose: **Ignore any differences in price level.**

You are incorrect.

This would be rather a rash approach. Tender price levels can vary quite dramatically over time and it is worth while making adjustments for these variations.

Although increases in prices are experienced most frequently, a fall in price level can also occur. This was the situation in the UK between 1989 and 1992, during a period of recession in the construction industry. At this time tender prices decreased despite increasing costs of labour, materials and plant. Contractors and specialist contractors had to price work very competitively to win projects.

Look at Fig. 9.1 which shows the cost indexes for both the UK and the USA between 1985 and 1996.

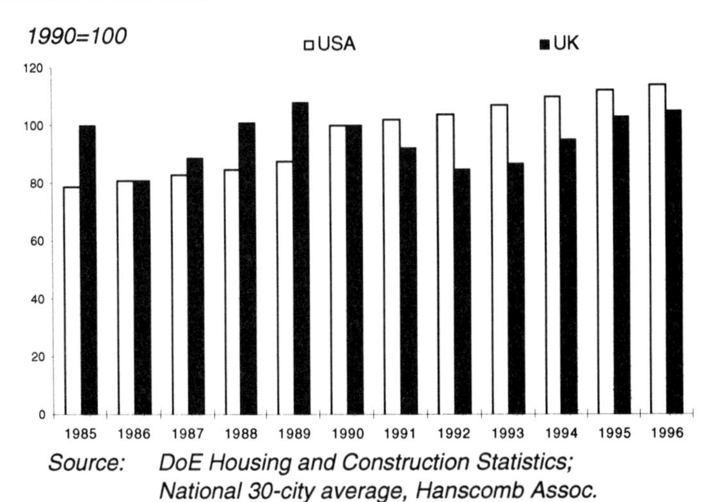

Fig. 9.1 US and UK construction cost indexes

Also, not to be ignored are the other important factors which affect Price level:

- difference in individual contractor's price levels relative to the general market price level (due to differences in allowances for overheads, profits, etc.);
- difference in regional trends, and localised variables;
- difference in tendering arrangements (including selection of contractors), contract conditions, contract period, weather conditions, site conditions, etc.

It is true that differences in cost due to the *other* Price level factors are usually smaller than those due to a change in the general market price level, but nevertheless an attempt should be made to allow for them.

Return to page 137

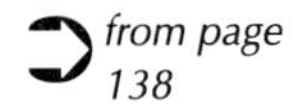
from page 138

Factor 3 Price level

Consider the hypothetical situation of two buildings which have been built from *exactly* the same detailed design. Any differences between the tender costs of these two buildings can obviously not be attributed to differences in Quantity and Quality. In this case the differences between the costs (i.e. total cost *and* cost of each element) are due to a difference in the **Price level**.

By 'Price level', we mean the price level of a project *relative* to general tender price levels at a specific date, and not *absolute* prices, which are, of course, primarily dependent on specification.

Price level includes several factors, all of which affect costs.

Write down at least three of the most important of the factors which collectively affect Price level by considering the hypothetical situation outlined above.

Turn to page 140

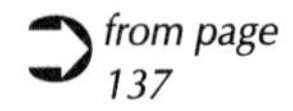
from page 137

You chose: **Construct and use an index which reflects changes in general tender price levels to clients.**

You are correct.

It is changes in tender levels to clients which are important and not changes in contractors' costs. Tender price levels can vary quite dramatically over time and it is worth while making adjustments for these variations.

Although increases in prices are experienced most frequently, a fall in price level can also happen, as was the case from 1989 to 1992 in the UK. This occurred even though contractors were experiencing increasing costs of labour, materials and plant. (See Fig. 9.1 on page 134.)

You have no need to construct your own index as BCIS monitors movements in price levels and publishes several price indices, including a Tender Price Index (TPI). However as the construction of the index relies on data received of live projects, the index will always lag behind the current date. Forecast figures are therefore needed to relate to the current date and future dates.

Also not to be ignored are the other important factors which affect Price level:

- ❑ difference in individual contractor's price levels relative to the general market price level (due to differences in allowances for overheads, profits, etc.);
- ❑ difference in regional trends, and localised variables;
- ❑ difference in tendering arrangements (including selection of contractors), contract conditions, contract period, weather conditions, site conditions, etc.

It is true that differences in cost due to the *other* Price level factors are usually smaller than those due to a change in the general market price level reflecting different tender dates, but nevertheless an attempt should be made to allow for them. In the UK the BCIS publish a range of price adjustment factors which allow tender prices to be modified with respect to location, contract size and procurement route. Professional judgement is required when assessing and making any adjustments for the remaining factors listed above.

For a more detailed description of:

BCIS tender price indices *Turn to page 141*

Other price adjustment factors *Turn to page 144*

If you would prefer to omit these sections at this stage, *Turn to page 146* from pages 134, 139, 140

To help in the assessing, the influence of (a) tender dates and (b) contractors' price levels, any analysis should provide the following information:

- **The *date of tender* so that the effects of any differences in the general market price level between two tender dates can be found by using an appropriate tender price index;**
- **The *location* for each project. By use of published location factors, projects in different locations can be brought to a comparable base;**
- **A comment on *market conditions* pertaining at the time of bidding.**

In addition, site conditions, tendering arrangements and contract conditions should all be specified on the front page of analyses, but it is left to the cost consultant using professional expertise and judgement to gauge their overall effect on costs.

Assume you wish to use the cost information included in a cost analysis of a project where the date of tender is three years previous to today's date.

Can you suggest a reasonable way of allowing for differences in Price level which have occurred over this time period?

Ignore any differences in Price level. Turn to page 134

Construct and use an index which reflects changes in general tender price levels to clients. Turn to page 136

Construct and use an index which reflects changes in contractors' costs due to increases in labour, materials, and plant charges. Turn to page 139

from page 133

The quality of each element can be assessed from the 'Specification and design notes' in the detailed analysis (Appendix A).

Another aspect of Quality is the value provided by the specification of an element. That is, the 'worth' of the element. If this aspect of Quality is not considered by the design team, the client may be involved in unnecessary expense. For instance, hand-made special joinery may be specified whereas standard joinery from a manufacturer's catalogue may equally suffice. The decision would need to be made whether the client is prepared to pay extra for the individuality of the joinery.

For this reason, a judgement of Quality based upon 'worth' might be more illuminating than a straightforward comparison of specifications.

Methods of allowing for differences in Quality based upon both these concepts will be demonstrated during the preparation of the cost plan in later chapters.

We shall now go on to discuss **Price level**.

Turn to page 135

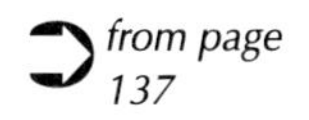
from page 137

You chose: **Construct and use an index which reflects changes in contractors' costs due to increases in labour, materials, and plant charges.**

You are incorrect.

It is prices to clients, as reflected in tenders received, which are important and not changes to contractors' costs due to increases in labour, materials, and plant charges.

Return to page 137

from page 135

The most important factors which affect Price level are:

- **(a) differences in general tender price level due to difference in tender date, reflecting inflation in labour and material rates;**
- **(b) difference in contractors' price levels relative to the general tender price level due to differences in allowances for overheads, profits, etc. when the market is low – then the contractors will be prepared to price more competitively to win work;**
- **(c) difference in regional trends, and localised variables;**
- **(d) difference in tendering arrangements (including selection of contractors for tender lists), contract conditions, contract period, weather conditions, site conditions, etc.**

Cost analyses must include as much relevant information as possible to enable the influence of each of these factors on costs to be assessed.

Examine Appendix A, then write down the information which BCIS *Detailed Cost Analyses* give to help assess the influence of factors (a) and (b) above.

Turn to page 137

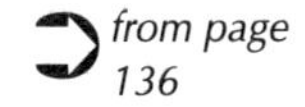
from page 136

BCIS tender price indices

What is a tender price index?

The purpose of a price index is to measure changes in price levels from one time period to another. Indices are published around the world. For the UK, tender price indices are published quarterly by the Building Cost Information Service (BCIS). These measure contractors' price levels for accepted tenders (i.e. cost to the client). All index figures are based on a given base date (expressed at a level of 100). Any changes in cost are related to this figure and can be used to calculate a percentage change over base year tender price levels.

The BCIS tender price index series is a statistical series which is based on a technique of sampling prices from the construction industry. Quarterly index figures use an analysis of projects with a contract sum in excess of £50 000. The average contract value of projects analysed is approximately £1 million.

How is it compiled?

Index calculations are based on the analysis of bills of quantities for about eighty projects. Each bill of quantities is re-priced using a standard base schedule of rates. The resulting *base* tender figure is compared with the actual tender figure in order to produce a *project index*. Published tender prices indices represent an average of all project indices included in the analysis. Individual project index figures are also adjusted in order to remove tendering differences due to project size, location and procurement route.

Types of index

Tender price indices (TPI) are produced in a variety of different formats. The most commonly used index is the all-in TPI. This covers new building work in the UK and reflects all sectors of the construction industry (public, private and housing). The index is based on a random sample of schemes and represents the general trends of tender prices across all sectors (see Fig. 9.2).

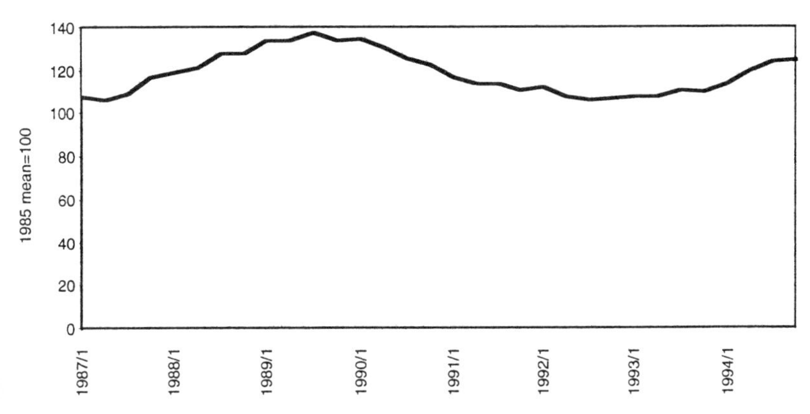

Fig. 9.2 All-in TPI

Turn to page 142

from page 141

The adjustment of tender prices to allow for time variations is carried out by applying the following formula:

$$A = \frac{(B - C) \times 100}{C}$$

where A = percentage change,
B = index at date required, and
C = index at date applicable to analysed prices.

Thus, the percentage change required to update a tender price, calculated during the first quarter of 1994 to an anticipated date of tender in the last quarter of 1994, can be expressed as:

$B = 126$ (index at last quarter 1994)
$C = 114$ (index at first quarter 1994)

$$A = \frac{(126 - 114) \times 100}{114} = 10.5\%$$

In other words, general tender prices rose by 10.5% during 1994.

Individual tender price indices are also produced for sub-sets of the 'all-in TPI'. These cover a variety of different types of construction projects and regions. The following categories are defined by the BCIS:

Public sector TPI This covers non-housing schemes for public sector clients including central and local government, health trusts, opted-out schools, etc. Some schemes for private sector clients are also included where funded from the public sector.

Housing TPI Based on a sample of housing schemes. The majority of schemes relate to social housing.

Private sector TPI Includes non-housing schemes for private sector clients. Some schemes for public sector clients are included where funded by the private sector.

Private commercial TPI A sub-set of the *private sector TPI* covering offices, shops, entertainment, garages, schools and colleges, agriculture, transport and health.

Private industrial TPI A sub-set of the *private sector TPI* including factories and warehouses.

Turn to page 143

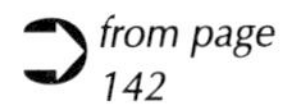
from page 142

Refurbishment TPI Based on a specially collected sample of schemes of work dealing with conversion and refurbishment throughout the UK and covering all sectors both public and private and including housing.

Housing refurbishment TPI A sub-set of the *refurbishment TPI* which covers housing schemes only.

Non-housing refurbishment TPI A sub-set of the *refurbishment TPI* excluding housing schemes.

Frequently, the sample size for a given category may be relatively small. In such cases the resulting index may be volatile. The BCIS recommends that these indices should not be used automatically for indexing of tender prices. Rather, they should be used to monitor any major divergence from general tender price movements (the all-in TPI). Figure 9.3 illustrates the difference between the all-in TPI and the private industrial TPI.

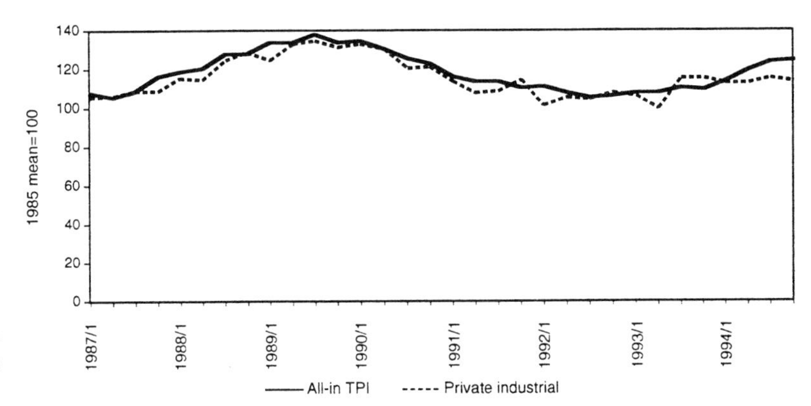

Fig. 9.3 TPI – all-in and private industrial

It can be seen that in the first quarter of 1992 the all-in TPI index was 112, whereas the private industrial index was 102. When considering projects relating to private industrial work, it may seem prudent to apply an index of 102.

However, in this situation the private industrial TPI *may* be unrepresentative of the true tender price level as the sample size used includes only a small number of projects. Thus, the all-in TPI *may* be more representative of likely tender prices due to the larger sample size. This illustrates the need for professional judgement when applying indices.

Turn to page 144

from page 143

In addition to producing historical tender price indices, forecast indices of likely future price levels are also published. These allow costs to be projected to a future tender date. Such forecasts are based on a range of published assumptions and economic indicators. Again, professional judgement will be required in assessing the likely accuracy of these indices.

Turn to page 172 for a brief discussion on the difference between a TPI and a cost index. (Make sure that you return to this page after reading the differences.)

from page 136

Other price adjustment factors

Location

The costs of a building are affected by a number of localised variables which give rise to a unique tender. These include local market conditions such as supply and demand for labour and materials, workload, taxation and grants. These variables are sufficiently volatile that not even identical buildings built at the same time within 10 kilometres of each other, obtain identical tenders.

In order to quantify the effects of location on tender prices for the UK, the BCIS publishes adjustment factors relating to Standard Statistical Regions, counties and local authority districts. The regions chosen are administrative areas and are not significant cost boundaries as far as the building industry is concerned.

While there will be small regional differences in the cost of materials due to transportation costs, the main influence is on the cost of labour. A lot of work in a particular region will increase the cost of the labour on a project.

Contract sum

Adjustment factors are also produced which allow the general relationship between tender prices and overall contract size to be incorporated into the analysis. The principle being that economies of scale are effective on larger projects, thus reducing the unit costs. There is also the principle of the learning curve effect where labour becomes more efficient when repetitive tasks are undertaken. For example, on high-rise buildings with repeat floors, the costs will reflect the repetition and efficiency.

Procurement route

Some variation in tender price can be expected due to differing forms of procurement. Adjustment factors are provided for competitive, negotiated and serial contracts.

Turn to page 145

from page 144

Price adjustment factors relate the impact of a given variable to the national average, for example, the cost implication of tender prices in Bristol compared with the national average for tender prices. Adjustment of tender price indices is carried out by simple multiplication, for example:

Index for first quarter 1995 = 126

Adjustment factors Location factor (Woking) = 1.10
Procurement factor (negotiation) = 1.12
Contract size factor (£4 724 000) = 0.90

$$\text{Adjusted TPI} = 126 \times 1.10 \times 1.12 \times 0.90 = 139.71$$

This adjusted TPI can now be used in any adjustment calculations. An example of this will be given in Chapter 10.

The statistical reliability of the price adjustment factors is dependent on two factors – the number of projects included in the sample and the variability of the factors. An average factor based on a small number of widely varying figures is likely to be less reliable than one based on a large number of closely grouped figures.

The difference between 126 and 139.71 is significant. Hence, care has to be taken when applying such adjustments to a cost plan. We keep referring to professional skill and judgement; in such a situation there is no substitute.

Turn to page 146

from pages 136, 145

Summary of Quality and Price level

An elemental cost analysis consists of an element-by-element allocation of costs.

The total cost of the building is expressed as 'cost per m^2 of gross floor area.' The cost of the building can also be expressed as a 'functional unit cost' (e.g. cost per m^2 of usable floor area, or cost per bed space, etc.).

The cost of each element is shown as 'cost per m^2 of gross floor area'. In addition, an 'element unit rate' is calculated for many, but not all, elements.

Sufficient details and specification notes are given of the factors which affect costs (Quantity, Quality, and Price level) allowing appropriate adjustments to be made.

Adjustments for Quantity have been considered in the previous section. Here we have been looking at the necessary adjustments for Quality and Price level and we saw that:

(a) For any one element, the element unit rates must be adjusted to account for variations due to Quality. Information necessary for this task can be gained by examining the **'specification and design notes'** for that element.

(b) Adjustments to account for variations due to Price level are also necessary and can be made by comparing the respective **tender dates, locations, site conditions**, and **contract conditions** given on the front page of the analyses. A **tender price index** and **location index** (such as those published by the BCIS) can then be used for part of the adjustments, but professional judgement is also required.

Turn to page 149

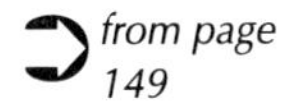
from page 149

The answer is **preliminaries and contingencies.**

Preliminaries and Contingencies

This chapter has been looking at factors affecting elemental costs, but as yet little has been said about **preliminaries and contingencies**. By referring to the analysis in Appendix A, you will see that a breakdown of the contract sum is given. Here the total for measured work is isolated from fixed sums such as provisional, prime cost and contingency sums. The total for preliminaries is also identified separately. This is the only place where the contingencies sum is declared (analysis of *element costs* excludes contingencies). The *Standard Form of Cost Analysis* recommends that both preliminaries and contingencies should each be expressed as a percentage of the remainder of the contract sum.

Preliminaries

These are items required to construct the project, which the contractor will price. They include such items as the cost of the site management team, scaffolding, site offices, toilets, canteen, first aid, insurances for the works, temporary protection, plant and equipment not specific to one task (such as dumper trucks, hoist and craneage), keeping roads clear of mud, removal of site debris, water and power for the works, temporary lighting, site telephones, cleaning the works, hoardings, safety requirements, and other general items specific to the contract conditions.

Contingencies

A contingency sum is included to cover unforeseen items and eventualities which occur during the construction of the project. It can vary in amount dependent upon the project size and type.

How are preliminaries priced by contractors?

The contractor's prices are based upon resources over time. In the case of a crane, the contractor will allow for the cost of bringing it to site, its erection, the weekly hire rate, the cost of the driver and the banksman, the fuel and insurances, the maintenance charge, the dismantling cost and the removal from site. The length of time the crane is required is crucial. From this information the weekly or daily rate will be calculated and the overall rate included in the preliminaries. The contractor has to ensure that, should the project be delayed and an extension of time granted, the additional cost incurred will be reimbursed. Similarly, the cost of the site management team, which may include a site manager, general foreman, site engineer, site quantity surveyor, safety officer and general labour, will all be priced according to the length of time they are on site.

Turn to page 148

from page 147

Under normal circumstances a contractor will consider the preliminaries and price only the items which have a cost consequence. Problems arise as there are a variety of methods for including these prices into the bills of quantities. These options are:

(a) to include each price against the appropriate item in the preliminaries section; or
(b) to show no individual priced items, but give a total only for the preliminaries section; or
(c) to spread prices attributable to preliminaries items evenly or unevenly throughout the measured rates in the bills of quantities; or
(d) to use various combinations of the above approaches.

This presents initial problems for the analysis of this element, and also later when comparisons are made with other projects. As far as the analysis is concerned, one way of overcoming the problem is to apportion any identifiable preliminary costs evenly throughout the other elements (see analysis in Appendix A, for example).

What other major problem exists with the analysis of the 'preliminaries' section?

Write down your answer.

Turn to page 150

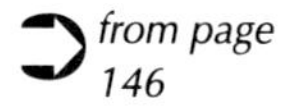
from page 146

From our discussions, it can be seen that it will be possible to alter and extend the diagrams on page 131. *'Element unit rate'* must now become *'adjusted element unit rate'* as shown below:

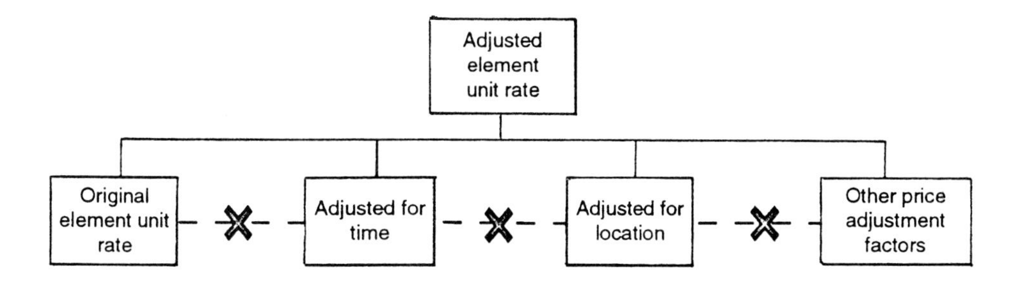

All the above adjustments may be modified after applying professional judgement.

There now remain two important factors which we have not mentioned yet.

Do you know what they are?

Write down your answer.

Turn to page 147

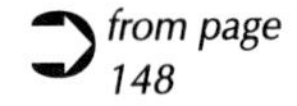
from page 148

A problem exists when trying to decide on suitable Quantities for this element.

We have talked about identifying one simple measure for each element, an **element unit quantity** (see page 127). Ideally this should provide a simple index of the relative size of each element. With preliminaries, it is generally accepted that this is best achieved by expressing preliminaries as a percentage of the remainder of the contract sum. In practice this is the normal approach, however research has shown that this is not a very satisfactory method. We will return to this subject again in later chapters when considering cost planning techniques.

Chapters 8 and 9 have briefly outlined how elemental cost analyses are set out. They have shown that analyses allow informed judgements to be made about the relative value for money offered by different buildings. This demands an understanding of costs and the factors which affect costs.

It is a short step from comparing the costs of buildings to using cost analyses to plan expenditure for future buildings, as shown in the following chapters.

Turn to page 151

10 Cost Planning during Feasibility, Outline Proposals, and Scheme Design

> This chapter *is* in programmed form and you must follow the page directions exactly.

This chapter deals with the first three stages in the RIBA *Outline Plan of Works*:

A Feasibility

B Outline Proposals

C Scheme Design

Turn to page 152

from page 151

Before considering the cost planning techniques used during Feasibility, Outline Proposals, and Scheme Design, let us revise what you learned during Part 1 of this programme. (*Note:* We have not included 'inception' as this does not involve cost planning.)

Which of the following statements correctly identifies a principle (or principles) of cost control common to all three stages of design mentioned above?

(1) There must be a frame of reference.
(2) There must be a method of checking.
(3) There must be a means of remedial action.
(4) There must be a frame of reference and a method of checking.

Write down your answer.

Turn to page 154

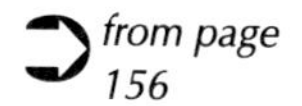
from page 156

The relationship between available information and the design and cost decisions made during each design stage is as shown below.

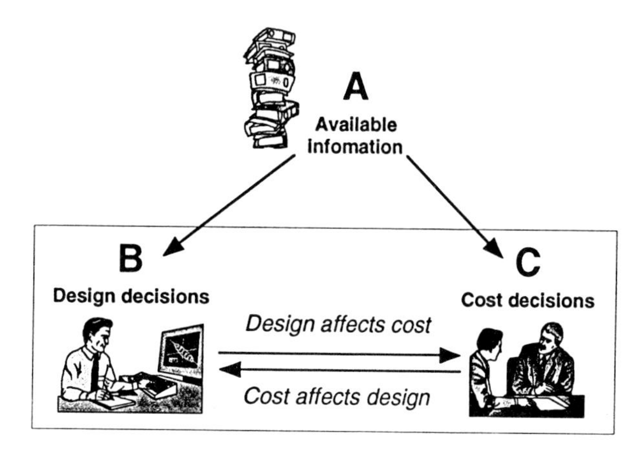

Study the table given on page 155 and the chart in Appendix B on page 320.

What items comprise A, B, and C at each of the design stages?

Write down your answer.

(If you find yourself getting lost while you are reading the remainder of the book, refer back to this short section, pages 152–156, to remind yourself of how the design process and cost planning techniques come together.)

Turn to page 155

from page 152

Based on the elementary view of cost control given in Part 1, the expected answer is 'There must be a frame of reference'. It is the *only* principle incorporated in all of the three stages of design mentioned, i.e. feasibility, outline proposals, *and* scheme design.

Principle	Stage
there must be a frame of reference	feasibility, outline proposals, scheme design
there must be a method of checking	detail design
there must be a means of remedial action	detail design

The feasibility, outline proposals and scheme design stages are all concerned with ***planning*** both the design and the cost of the construction. The detail design stage is the crucial stage in ***controlling*** the cost of the design. Remedial action follows cost checking and also takes place during the detail design stage.

In one sense, however, you are correct if you said that there must be a method of checking during outline proposals and scheme design. Control would be lost if there was no way of checking that the cost limit had not been exceeded during the preparation of both the outline cost plan and the detailed cost plan.

In other words, there must be a 'method of checking' (and a 'means of remedial action') during the preparation of the 'frame of reference'.

Turn to page 156

from page 153

		Feasibility	Outline Proposals	Scheme Design
A	*Information available at the beginning of the stage*	Initial brief Space requirement Functional use of project Quality expectations Site location and planning constraints	General brief Planning policy Feasibility report *(including the agreed estimate)* Design criteria showing shape, overall layout and general arrangement	Detailed brief Outline design Outline cost plan
			↓ *used for*	
B	*Design report at the end of the stage comprises . . .*	Statement of technical feasibility	Architectural, structural and engineering service concepts Outline design drawings	Brief specification Scheme design drawings
			↓ *and*	
C	*The cost report which comprises . . .*	First estimate	Outline cost plan	Detailed cost plan

Turn to page 157

from page 154

The 'frame of reference' used for cost control is prepared during the feasibility, outline proposals, and scheme design stages; both cost checking and remedial action (if necessary) take place during the detail design stage.

As you learned previously, the preparation of the frame of reference consists of:

- ❑ establishing a realistic first estimate (during feasibility), and
- ❑ planning how this estimate should be spent among the various parts of the building (during outline proposals and scheme design).

This planning is carried out during two stages:

(1) **Outline proposals** A cost target is prepared for *each major group of elements;*
(2) **Scheme design** A cost target is prepared for *each element.*

It should become obvious as you read through the remainder of the book that:

- ❑ the cost decisions made during feasibility are based on the client's functional requirements rather than on a particular design solution,
- ❑ the decisions made always take account of the limited nature of the information upon which they are based,
- ❑ the decisions are always made in such a way as to impose minimal restriction on the range of possible design solutions later on.

Turn to page 153

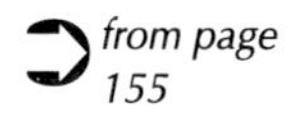
from page 155

Section A

Cost planning during feasibility

Section A
Cost planning during feasibility (page 157)

Section B
Cost planning during outline proposals (page 187)

Section C
Cost planning during scheme design (page 202)

Turn to page 158

from page 157

Design teams are usually required to produce a first estimate at the feasibility stage *long* before any drawings have been prepared or any specification decided. Because of this, traditional estimates based on approximate quantities are necessarily approximate, and are not to be relied upon.

The term 'approximate'

It is unfortunate that the term ***approximate*** *has become accepted terminology. The approximation reflects the lack of detailed information available about the project in the early stages. Even during the scheme design and the detail design stages the approximate quantities approach is still used. Only when design information is firm will the term be dropped.*

Despite this fact, clients expect to be able to make the decision to proceed, modify and proceed, or abandon the project at the end of the feasibility stage. After all, *feasibility* means just that.

Thus, clients need to have relevant and reliable cost information together with confidence in the design team's ability to manage costs. This will reduce the risk of difficulties later in the design process.

Design teams must therefore adopt techniques for making first estimates more reliable. These techniques must obviously be based on all the available information at this stage.

What information is usually available at the feasibility stage?

Sketch plans. Turn to page 160

Initial brief. Turn to page 163

Both of these. Turn to page 160

from page 163

The initial brief will contain usually only outline description of:

(a) the use to which the building is to be put (e.g. hotel, school, office),
(b) the usable floor area required (additions will thus have to be made to allow for circulation areas, etc.),
(c) an indication of the quality required (e.g. for a proposed hotel whether it is to be 'one star' or 'four star') – reference is sometimes made to similar existing buildings,
(d) details of the site.

We now consider the **interpolation method** of preparing a first estimate.

Under this method, costs of other buildings are studied, bearing in mind the space and standards provided by these buildings. Therefore, information on suitable projects needs to be assembled and studied. Efficient organisations will make use of computers to store their cost data files (BCIS is a good example). The chief advantage of such an approach is that large quantities of data can be stored and rapidly manipulated, which allows a wider range of cost studies to be considered than might be the case with manual methods. Various characteristics of the data need to be referred to when selecting appropriate cost studies for use.

For selection purposes, while all the items are important, which of the following criteria do you consider to be the *most* important?

(a) Only the costs of buildings of approximately the same gross floor area as the proposed building. Turn to page 164

(b) Only the costs of buildings where the site conditions are nearly the same as the proposed building. Turn to page 161

(c) Only the costs of buildings of the same general use (i.e. the same type of building) as the proposed building. Turn to page 165

from page 158

You chose either **Sketch plans** or **Both of these.**

You are incorrect.

Sketch plans are not usually produced until the outline proposals stage (see Appendix B), *unless the client provides sketch plans or the architect starts drawings before the economic feasibility has been considered.*

We suggest that reliable estimates can be prepared when the only available information about a new project is the initial brief. Not everyone will agree with this proposition. We admit that there will be cases where the design process must be taken to a more advanced stage before a reliable estimate can be given. However, this should not occur very often. If you would like to see how we suggest that such an estimate should be prepared, turn to page 229. When you have read through the preparation of the first estimate given between pages 229 and 242, turn back to this page.

Is the initial brief usually available at the beginning of the feasibility stage?

Write down your answer.

Turn to page 163

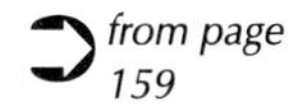
from page 159

You are incorrect. This is not the most important criterion.

Even *major* differences in site conditions can usually be allowed for more easily than the fundamental differences between two types of building such as schools and factories. Approximate quantities could be used, for instance, to make allowance for differences in site conditions.

Do you consider it more important to isolate analyses of buildings of the same type or analyses of buildings of approximately the same gross floor area as the proposed building?

Buildings of the same type. *Turn to page 165*

Buildings of approximately the same gross floor area. *Turn to page 164*

from page 164

Notice that the wall-to-floor area ratio and the cost of walls per m^2 of gross floor area only decreases by a factor of *10* whereas the gross floor area has increased by a factor of *100*.

The example on page 164 is rather impractical as, generally speaking, only buildings such as warehouses would be built 50 metres square. It would, therefore, be more realistic to consider how a single-storey building of length 500 metres and width 5 metres compares with building A. You might say that this also is an unrealistic size. Nevertheless, we will use it for our example. Thus, compared to building A, building B comprises:

Description	
Plan area	500 × 5 m
Gross floor area	2500 m^2
External wall area	3030 m^2
Wall-to-floor area ratio	1 : 21
Total cost of walls	£757 500
Cost of walls per m^2 of gross floor area	£303

Hence, while the increase in floor area is the same as for building B (i.e. by a factor of 100), the wall-to-floor area ratio and cost of walls, per m^2 of gross floor area, have in this example approximately halved when compared with building A.

This is why we say on page 164 that only *very* large differences in gross floor area (e.g. by a factor of 10 or more*) will not be catered for by considering costs per m^2 of gross floor area.

You should have selected 'Buildings of the same type', as being the most important criterion.

Turn to page 165

* This is given as an example and should not be taken as a general rule, as it will vary with every building type and size.

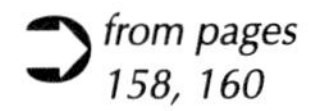
from pages 158, 160

You chose **Initial brief** or wrote down **Yes.**

You are correct.

The initial brief is the only information available at this stage (see page 155). To appreciate the limited nature of information contained in the initial brief, consider the following.

Initial brief for the office block discussed in Chapter 11

Space	A lettable area of at least 1850 m^2 and maximum permissible car parking. The parking provision is considered to be essential due to the congested city centre site, poor parking facilities and inadequate public transport. The tenants are likely to be professional services, such as accounting and insurances.
Planning permission	Preliminary enquiries to the planning authority indicate that an office block three storeys high would be permitted and that the stated amount of lettable area should be achievable as it is within the plot ratio.
Use	The premises will be offered for leasing as good quality office accommodation at a rental which is in the upper market range.
Quality	An above average commercial quality is appropriate to this prime location site and will command a suitable return. Raised floors and suspended ceilings will need to be incorporated into the new design in order to accommodate IT systems. An attractive façade is considered essential.
Site	The site has been acquired and a site plan is available. (It has not been included here as it is reasonably straightforward.) It shows that the terrace houses previously occupying the site have been cleared and the foundations grubbed up. Ground conditions in the area are good.

How would you summarise the types of information that are usually contained in an initial brief?

Write down your answer.

Turn to page 159

from pages 159, 161, 165, 166

You are incorrect. This is not the most important criterion.

Differences in gross floor area can usually be allowed for by considering costs per m^2 of gross floor area. Only *very* large differences in gross floor area (e.g. by a factor of 10 or more*) will not be catered for by this method. Where a difference in floor areas does occur between two buildings it is necessary to take account of the decrease in the wall-to-floor area ratio which will result. *This gives rise to a decrease in the costs of external walls, windows, etc., per m^2 of floor area.*

Example

	Analysed building A	Analysed building B
Dimensions		
plan area height	5m × 5m, 3m	50m × 50m, 3m
gross floor area	25 m^2	2500 m^2
external wall area	60 m^2	600 m^2
Wall to floor ratio		
$\frac{\text{wall area}}{\text{gross floor area}}$	2.4	0.24
Assume element unit rates for external walls are:	£250 per m^2	£250 per m^2
Total cost of walls	£15 000	£150 000
Cost of walls per m^2 of gross floor area	£600	£60

Note that the £150 000 and the £60 per m^2 gross floor area for the proposed building could be calculated if the information for building A, listed above, was given, so even *large* differences in gross floor area can be allowed for.

Turn to page 162

* This is given purely as an example and should not be taken as a general rule, as it will vary with building type and size.

from pages 159, 161, 162

You are correct. 'Buildings of the same type' is the most important criterion.

Only buildings of the same general functional use should be examined. Any variation in site conditions can be allowed for relatively easily, while variations in floor area can be adjusted by considering costs per m^2 of gross floor area. Only *very* large differences in gross floor area (e.g. perhaps by a factor of 10 or more) will not be catered for by this method. (If you wish, see page 164 for a discussion of this.)

At this stage it is important to assemble a range of costs for the particular building type required. This can be done by referring to cost analyses contained in office records, or from the BCIS, or from analyses published in the technical press.

For example, the range of costs for office blocks, used in Chapter 11 (pages 230–1) is typical of information published by BCIS. The details are presented in the form of a histogram and costs are expressed in £ per m^2 of gfa. From such information it is possible to gain *some idea* of the correct cost range. However, accuracy can be improved and more details concerning quality can be considered if detailed cost analyses are used.

Research has indicated that it is a mistake to reduce the number of cost analyses used too far. For example, you would be ill advised to use just one analysed project. Even given the use of computers, too many analyses can also present problems of a practical nature. A sensible approach suggests that the use of between 4 and 10 analysed projects is a suitable compromise. (It should be emphasised here that it is dangerous to take one element from one analysis and a second element from another analysis.)

The estimate is prepared by interpolating on the range of costs of buildings of the same use, rather than simply choosing one of the costs unaltered.

Interpolation on the range takes place in four stages:

(1) choosing the most suitable number of cost analyses (say between 4 and 10);
(2) isolating the major differences between the buildings analysed and the proposed building;
(3) making allowances for these differences, and
(4) then applying the interpolation using professional judgement combined with simple statistical procedures, such as measures of central tendency, measures of dispersion, etc.

Which of the following items of information from a brief is it most important to bear in mind when making the *final* choice from the most suitable cost analyses?

Space. Turn to page 166

Use. Turn to page 167

Quality. Turn to page 169

from pages 165, 167

You are incorrect.

Small variations in gross floor area can be allowed for by considering costs per m^2 of gross floor area. The cost consequences of very large differences in gross floor area may need further consideration. (On page 164, it would be possible to calculate the figure of £150 000 (or £60 per m^2 of gross floor area) if all other information was available.)

Since it is more difficult to allow for differences in *use* or *quality*, by choosing *space* you have chosen the least important of the criteria!

When initially selecting suitable analyses for the preparation of a first estimate, choice will centre on buildings of the same type (and use) as the proposed building.

We asked you to identify the most important item with regard to making the final choice. You should have selected 'Quality'.

Return to page 165

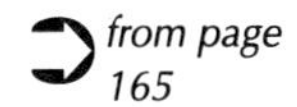
from page 165

You are incorrect.

When initially selecting suitable analyses for the preparation of a first estimate, choice will centre on buildings of the same type (and use) as the proposed building. We asked you to identify the ***most important item*** with regard to making the ***final*** choice, so choose again:

	Most important item
Selection of range of costs	of similar *use*
Final choice from the range	of similar...

Which is the correct word to complete the table above?

Space. Turn to page 166

Quality. Turn to page 169

from page 169

Remember, ***all*** information available should be used during the preparation of the first estimate. Valuable information is often contained in the main part of the cost analysis.

The major differences which are likely to come to light are:

(a) difference between the date of tender of the analysed building and the date of preparation of the first estimate (Price level);
(b) a difference in location (Price level);
(c) a difference in the gross floor area (Quantity);
(d) items in the analysed building which will not be required in the new project (mainly Quantity but perhaps Quality also);
(e) items which are missing from the analysed building but which will probably be needed in the proposed building (mainly Quantity but perhaps Quality also);
(f) significantly different requirements in the preliminaries and contingencies elements, e.g. temporary roads or temporary works.

Notice that the statement of Quality in an initial brief is usually vague. In fact, relatively small differences in quality which remain after choosing any cost analyses cannot be allowed for until outline proposals or later.

It is highly likely that the costs contained within all the chosen analyses will reflect different tender dates and locations. Major gains towards comparability can often be made if price level is considered first. Therefore, adjustments will be required to bring the price level of the analyses in line with the requirements of the new project.

Write down a *brief* statement of how to allow for the difference in the general tender price level between the dates of tender of the buildings analyses and the date of preparation of the first estimate.

Turn to page 172

from pages 165, 166, 167

You are correct.

The usual final procedure is to select the cost analyses of buildings which approximate most closely to the **quality** desired by the client.

Thus the choice of cost analyses is made in two stages:

Stage 1 — A range of cost analyses is assembled of the same ***type*** of building as the proposed building.

Stage 2 — This range is examined to find between 4 and 10 buildings whose ***quality*** approximates most closely to that required by the client.

For many organisations the first stage is a normal on-going exercise of data collection, analysis, classification and storage. Where in-house data is lacking, then other sources, such as the BCIS, can be used as a supplement.

The efficiency of the second stage will depend partially on whether manual or computer-assisted methods are used. The quality and extent of the data sources, together with the approach and time available, will influence the suitability and number of analyses finally chosen.

When enough suitable analyses have been selected, then the differences between the chosen analyses and the proposed building are isolated. Allowances are made for these differences. The estimate for the proposed building is completed using this adjusted cost information and professional judgement.

When isolating the major differences between the chosen analyses and the proposed building, what should be studied?

(a) The brief of the proposed building and the information contained on *page 1* of each BCIS analysis (or its equivalent).

(b) The brief of the proposed building and *all* the information contained in each BCIS analysis (or its equivalent).

(If necessary, refer to the analysis in Appendix A.)

Turn to page 168

from page 176

So far we have considered the selection of a narrow range (between 4 and 10) of suitable analyses.

We then discussed adjustments to these analyses to take account for differences in time and location.

These analyses will reflect buildings of varying size and we must now examine how these differences are allowed for. From Chapter 8, you will remember that the common yardstick is the cost per m^2 of gross floor area. This rate is used to compare the costs of buildings of different gross floor areas.

An overall cost per m^2 of gross floor area is selected from the range of cost analyses. This will be determined by interpolation using professional judgement, aided perhaps by simple statistical measures (e.g. measures of central tendency and measures of dispersion).

Our selected cost per m^2 of gross floor area is simply multiplied by the gross floor area required in the proposed building. Complications can arise when other allowances are necessary. This will be discussed in the next few pages.

One complication occurs if there is a *very* large difference between the size of the project and any cost analysis. It may be necessary to allow for the cost effects of a reduction in the wall-to-floor area ratio which usually accompanies a large increase in gross floor area (as discussed on page 164).

We now turn our attention to adjustments for major differences between the analysed buildings and the proposed building.

Turn to page 177

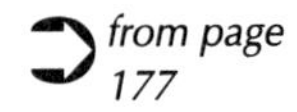
from page 177

You are incorrect.

You are wrong on *two* counts:

- you suggest subtracting a 1992 price from a 1995 price – this would obviously give a meaningless cost;
- you suggest subtracting a lump sum appropriate to a building of one gross floor area from a lump sum appropriate to a building of a different gross floor area (the two areas are 1408 m^2 and 2250 m^2).

To make this deduction from the updated total cost of the proposed building, the £14 040 would first have to be updated

$$£14\,040 \times \frac{128}{106}$$

and converted to a cost appropriate to the proposed building

$$£14\,040 \times \frac{128}{106} \times \frac{2250}{1408}$$

Return to page 177 and choose again

from page 168

Any difference in general tender price level and the date of preparing the first estimate is allowed for *by using an appropriate index* (see Chapter 9).

It is important to distinguish between the various types of indices which are available. Two of the most common are 'factor cost indices' and a 'tender price indices'.

- ❑ A **factor cost index** is an input index that measures changes over time in resource costs such as labour, material, and plant (i.e. costs incurred by contractors).
- ❑ A **tender price index (TPI)** is an output index that measures changes in tender price (i.e. the price a client has to pay for its building).

A 'tender price index' is therefore more appropriate for our purposes here.

Both indices can move independently. In the early 1990s, building costs continued to rise gradually while tender prices were falling. This trend was a reflection of the severe recession and lack of orders for new work affecting the building industry.

This shows that although contractors were paying higher prices for materials and equipment, they were absorbing the higher costs by increases in productivity or by lower profit margins.

In the worked example in Chapter 11, the date of tender in the analysis is September 1992 and the first estimate for the 'proposed building' is dated June 1995.

Example

Date	Tender price index	Factor cost index
September 1992	106	146
June 1995	128	162

When this first estimate is being prepared, how should the costs from the analysis be updated to allow for the increase in the general tender price level which has occurred?

Write down your answer.

Turn to page 174

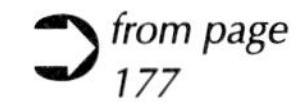
from page 177

You are correct.

Of the three possibilities, this is the only completely correct one. You may, of course, prefer to subtract £14 040 from the total cost of the analysed building, convert the result to a cost per m^2 of gross floor area of the analysed building, then update. The arithmetic is equally complex in both cases, so it does not really matter which one is used.

The main points to notice about deductions are:

(a) they should either be made before updating any costs, or after updating ***all*** costs; whichever method is used care must be exercised to ensure that costs are compatible;
(b) they should be made before the total cost per m^2 of gross floor area is multiplied by the gross floor area of the proposed building.

We shall now discuss the slightly more complicated problem of allowing for items in a proposed building which have not been included in the analysed building.

Turn to page 178

from page 172

From the indices on page 172,

$$\text{building costs in June 1995} = \text{costs in September 1992} \times \frac{128}{106}$$

(remember, by 'building costs' we mean costs to the client i.e. the builder's 'prices').

Costs from the analysis may therefore be updated to June 1995 by multiplying by the factor

$$\frac{128}{106}$$

(This ratio can be converted and shown as an Adjustment Factor (AF). Thus here, the AF would be 1.21.)

Notice that updating will always involve multiplying by

$$\frac{\text{the index at the later date}}{\text{the index at the earlier date}}$$

This adjustment can also be expressed as a percentage:

$$\frac{128 - 106}{106} \times 100 = 20.75\%$$

For updating purposes it is easier and quicker to use the ratio method rather than calculating and making any adjustments via a percentage.

When the price level has been rising, will this ratio of indices be greater or less than 1?

Write down your answer.

Turn to page 176

from page 178

Allowing for additions becomes more complicated when:

- **it is impossible to obtain a realistic cost per m^2 of gross floor area for an additional item;**
- **when it is not known exactly how much (what quantity) of the additional item will be required.**

Example

From Appendix A we find that the carpet tiles for the floor finishes were supplied by the client. The cost of supplying carpet tiles has to be added to the first estimate prepared in Chapter 11.

Which of the following methods do you prefer?

(a) **Find a current cost per m^2 of gross floor area for the supply of carpet tiles from a manufacturer's catalogue and multiply it by an assumed area of carpet tiles required for the proposed building.**

Turn to page 182

(b) **Multiply a current rate for the supply of carpet tiles by the area of carpet tiles in the building analysed. Divide this by the gross floor area in the analysis and add the result to the updated cost per m^2 of gross floor area. Finally, multiply this rate by the gross floor area of the new project.**

Turn to page 180

from page 174

When the price level has been rising, the index will have increased, so the ratio of indices will be *greater* than 1.

Updating for time is not the only item which needs to be considered. Other factors include location, selection of contractor, form of contract, type of construction, etc.

Of these items, location is perhaps the next in order of importance. Any location adjustment is relatively straightforward as any difference in price level due to location can be allowed for ***by using an appropriate factor index***, which is adjusted in much the same way as that for the time allowance.

Example

Location	BCIS location factors (September 1995)
Analysis – *Reading*	1.00
New project – *Portsmouth*	0.94

This means that, on average,

$$\frac{\text{new project location}}{\text{analysis location}} = \frac{0.94}{1.00}$$

Costs from the analysis should therefore be adjusted for the new location by multiplying by this ratio.

(This ratio can be converted and shown as an Adjustment Factor (*AF*). Thus here, the *AF* would be 0.94. Furthermore the time and location adjustments can be combined to give a single adjustment factor.)

Turn to page 170

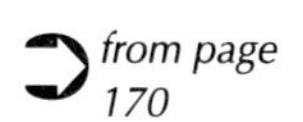
from page 170

The following example from Chapter 11 illustrates the method of deduction of items in a cost analysis which are not required in a proposed building.

Example

The cost of demountable office partitioning is included in the analysis (Appendix A), whereas in the new project this will be the tenant's responsibility. This means that deductions will have to be made for the partitioning. Let us consider this deduction. From the 'Specification and design notes' of Appendix A, we find that the cost of the demountable partitioning was £14 040.

Which of the following methods could be used for making the deduction?

(a) **Multiply the updated total cost per m^2 of the building analysed by the gross floor area of the proposed building, then subtract £14 040.**

Turn to page 171

(b) **Convert £14 040 to a cost per m^2 of gross floor area of the building analysed, subtract this from the total cost per m^2 of gross floor area of the analysed building, update the result, and finally, multiply this rate by the gfa of the proposed building.**

Turn to page 173

(c) **Convert £14 040 to a cost per m^2 of gross floor area of the building analysed, then subtract the result from the updated total cost per m^2 of gross floor area of the building analysed, and finally, multiply this rate by the gfa of the proposed building.**

Turn to page 179

from page 173

Allowance for additional items can be made if an appropriate cost per m^2 of gross floor area is obtained from another cost study or analysis. This need only be updated (and adjusted for location, etc., if necessary), and then multiplied by the gross floor area of the proposed building.

Example

Let us assume that a fire protection sprinkler system is required for the new project but no such provision is found in any of the selected cost analyses.

It is ascertained from other cost studies and from specialist installers that an allowance of £14.00 per m^2 should be adequate for a sprinkler system in the proposed office block. This is added to the total allowance per m^2 for the proposed building.

Notice that, a reasonably reliable allowance can be made even though the *exact specification (size and quality)* of the item has not been decided.

Write down a simple description of a situation in which it would be more difficult to make an allowance for an additional item.

Turn to page 175

from page 177

You are wrong in principle because the £14 040 is never updated.

This is because the total cost per m^2 of gross floor area of the analysed building is updated before the cost of partitions is subtracted. However, the figure obtained by this method would not be far wrong.

A more realistic figure would be obtained if the subtraction was performed either before updating any costs or after updating both the total cost and the cost of the demountable partitions.

Turn back to page 177

from pages
175, 182

You have chosen the best and most likely method. However, even this method is not likely to be very accurate but it will give a more realistic answer than guessing the specific area of carpet tiles which will be required in the new project.

In case you are not sure how this allowance would be made, here are the details:

(1) The current cost of carpet tiles per m^2 of gross floor area in the building analysed would be

$$\frac{\text{(current rate for supply of carpet tiles)} \times \text{(area of carpet tiles in building analysed)}}{\text{total floor area of building analysed}}$$

(2) This figure would then be added to the updated total cost per m^2 of gross floor area, and finally it would be multiplied by the gross floor area of the new project.
(This assumes that the proportion of the gross floor area covered by carpet tiles in the proposed building will be the same as in the building analysed.)

If you would like to read through a summary of the sources of cost information most commonly used for making allowances for additional items in new projects,

Turn to page 181 

If you prefer to continue, Turn to page 184 

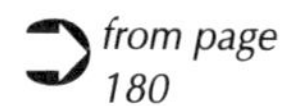
from page 180

Reference to other analyses

The costs and quality of the required item are compared in several cost analyses. An appropriate cost per m^2 of gross floor area is then chosen for the proposed building.

Use of published or private cost studies

Regular articles can be found in all the technical journals. Also, various trade associations publish cost studies from time to time. Other sources are the BCIS and government departments.

Reference to published price books

Due to trade discounts, reference to catalogues is never as good as telephoning suppliers direct. In practice, reference to published price books is used quite frequently, but it has its dangers. One problem is the vast difference in prices found in different price books for supposedly similar items.

Approximate quantities

The approximate quantities approach should be avoided wherever possible for the reasons discussed in Part 1. However, it is sometimes necessary to develop hypothetical design solutions for particular items, measure approximate quantities, and price the items. This method had to be used in Chapter 11 for the car parking and the external works.

Estimates of prices and rates from contractors and specialist contractors

For a specialist item, such as a sprinkler system, it would be advisable to contact a specialist installer to discuss the specification, the installation requirements and to obtain guide prices. Specialist contractors will frequently provide such prices and, while they will not accept any liability for the prices given, their aim is to secure work by providing such a service.

Turn to page 184

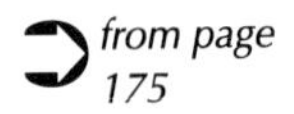
from page 175

It is very unlikely that there will be sufficient information available during feasibility to give an assumed area of carpet tiles required for the proposed building.

It would be more realistic to use the limited information which is available: namely, the proportion of the gross floor area which is covered by carpet tiles in the building analysed. The only way that this information could be used is, of course, to assume that the proportion of the gross floor area covered by carpet tiles in the new project will be the same as in the building analysed.

To see how this allowance is made ... *Turn to page 180* 

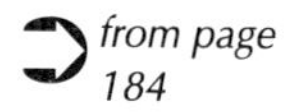
from page 184

Copy down the following description of the interpolation method, then fill in the blanks in your copy.

When you have done this, compare your description with the one on page 185.

Preparation of first estimate by the interpolation method

(1) A range of ____ ________ of the same ____ of building as the proposed building is assembled.
(2) This range is examined to find the ____ ________ (say between 4 and 10) whose ______ approximates most closely to that desired by the client.
(3) Adjustments for both ____ and ________ are made to bring all cost analyses to a current cost base appropriate for the location of the new project.
(4) The initial brief of the proposed building and ___ of the information given in the cost analyses are studied in order to ______ the _____ __________ between the proposed and existing buildings.
(5) An initial estimate of ________ _____ is calculated from the analyses, using the gross floor area of the new project, and then
(6) _________ are made for each of the _____ __________ (including, of course, site works, drainage and external works).
(7) The first estimate is built-up using ___________ ________ aided perhaps by simple statistical measures, depending on how many cost analyses are to be considered.
(8) An addition is made for ____________.
(9) Next, an allocation is included for ___________ which is intended to deal with unforeseen events or circumstances which might arise during the construction stage.
(10) Finally, an allowance is made as a reserve against price and design risk. It is intended that the allowance for price risk should offset rises between the general tender price level at the date of __________ of the _____ ________ and the __________ price level on the tender, whereas the design risk allowance is intended to provide a buffer against _________ ______ ___________ which may only come to light later in the design process.

Turn to page 185

from pages 180, 181

We are essentially considering design. However, attention should be paid to client instructions before reporting any estimate. What was asked for? There can be several options. For example, construction costs at proposed tender date; construction costs at completion; total budget of construction costs, including professional fees and value added tax on both construction costs and fees; etc. Be sure to address clients' requirements and to clarify any reported figures, stating clearly both inclusions in and exclusions from the estimate.

With this in mind, remember that when all the major differences between the buildings analysed and the proposed building have been allowed for, this first estimate is still not complete. Firstly **preliminaries** should be considered. It needs to be established whether any allowance has been included for preliminaries within the figures derived from the previous cost analyses. If not, an addition should be included at this point.

Contingencies are usually excluded from cost analyses. An allowance for this sum now needs to be inserted. The contingency sum is intended to cover those unforeseen events which have a tendency to arise during the construction phase of a project. Although this has little relevance to the design stages, contingencies must be included now for client budget purposes. In Chapter 11 we have allowed 1.5%; this percentage, we judge, reflects the type of project and the risks involved.

Next, consideration should be given to the likely increase in general tender price level which might be anticipated between the *feasibility* stage and the *receipt of tender* for the new building. This can usually be done by reference to a tender price index forecast (or similar forecast figures). It has to be stressed that these only forecast general trends and are not infallible. Professional judgement, including knowledge of local conditions, needs to be applied when interpreting any forecast figures. In Chapter 11, we judge the ***reserve against price rises*** is 5% of the remainder of the first estimate. A 'price risk' allowance is also included in the outline cost plan at outline proposals and the detailed cost plan at scheme design. The size of the allowance is calculated afresh at each stage using the latest information available.

Finally, to accommodate some flexibility in the design process, an allowance for design risk can be contemplated. Some clients and design teams hold the view that a separate allowance should not be added for design risk. Rather they see that a design risk allowance is included in the feasibility estimate. Under these circumstances, the reserve for design risk would need to be identified, and shown separately, during production of the outline cost plan. In Chapter 11, we have chosen to indicate an allowance for design risk at feasibility stage.

Without defining *how* allowances are made, write a short account of the basic steps in preparing a first estimate by interpolation.

(Refer back to pages 165 and 169 if necessary.)

Turn to page 183

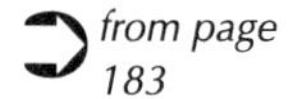
from page 183

Summary – Preparation of first estimate by 'interpolation'

(1) A range of **cost analyses** of the same **type** of building as the proposed building is assembled.
(2) This range is examined to find the **cost analyses** (say between 4 and 10) whose **quality** approximates most closely to that desired by the client.
(3) Adjustments for both **time** and **location** are made to bring all cost analyses to a current cost base appropriate for the location of the new project.
(4) The initial brief of the proposed building and **all** of the information given in the cost analyses are studied in order to **isolate** the **major differences** between the proposed and existing buildings.
(5) An initial estimate of **building costs** is calculated from the analyses, using the gross floor area of the new project, and then
(6) **Allowances** are made for each of the **major differences** (including, of course, site works, drainage and external works).
(7) The first estimate is built-up using **professional judgement** aided perhaps by simple statistical measures, depending on how many cost analyses are to be considered.
(8) An addition is made for **preliminaries**.
(9) Next, an allocation is included for **contingencies** which is intended to deal with unforeseen events or circumstances which might arise during the construction stage.
(10) Finally, an allowance is made as a reserve against price and design risk. It is intended that the allowance for price risk should offset rises between the general tender price level at the date of **preparation** of the **first estimate** and the **contractor's** price level on the tender, whereas the design risk allowance is intended to provide a buffer against **unforeseen design difficulties** which may only come to light later in the design process.

(This preparation is shown graphically on the next page.)

The first estimate will form part of the design team's report on the feasibility of the project as a whole. If the client accepts this first estimate this will be regarded as the **cost limit** for the project.

In most cases the first estimate can be given as a lump sum. However, there will be situations **where the brief is insufficiently developed** or technical difficulties are so great that a firm estimate cannot be given until after the completion of an outline design. Here, the design team may decide that the best approach at feasibility stage is to provide the client with the **cost range** within which they are confident of producing a design.

Turn to page 186

from page 185

Preparation of first estimate by the interpolation method

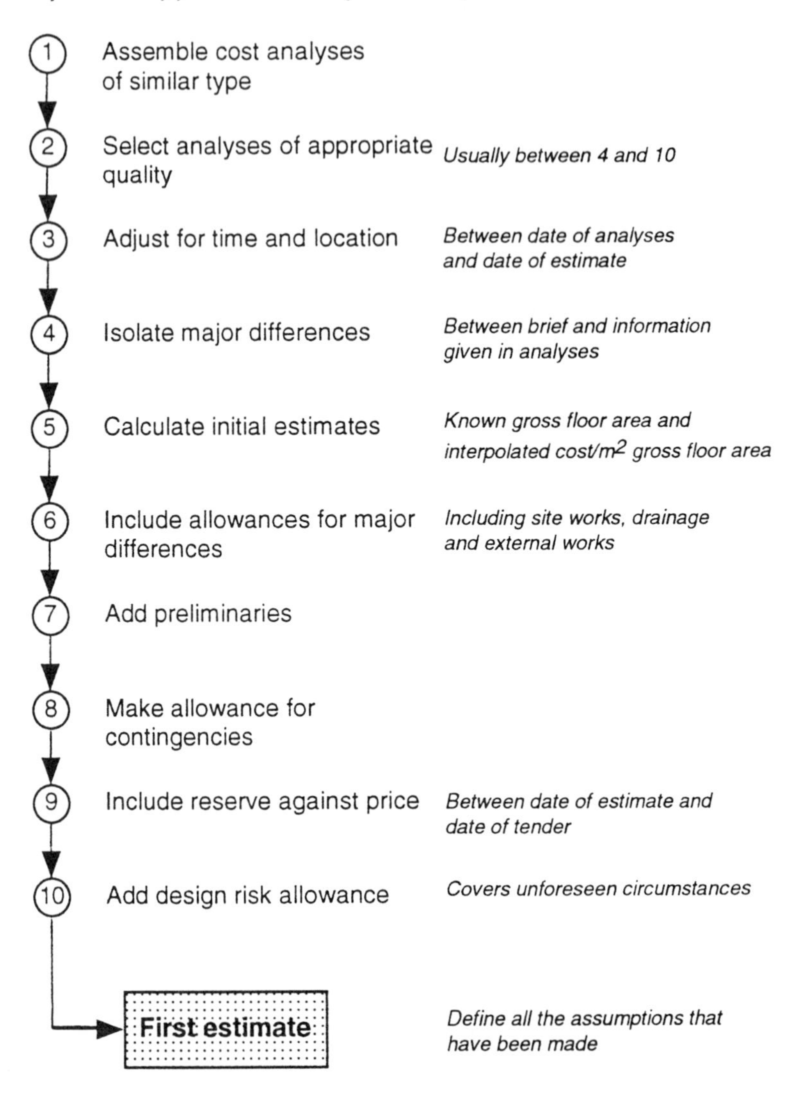

If you would like to read through an example of a first estimate being prepared by 'interpolation', read pages 228 to 242, then return to this page.

If you will have to stop soon, stop now. The next section of this chapter deals with cost planning during outline proposals.

Turn to page 187

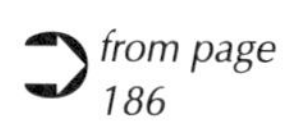
from page 186

Section B

Cost planning during outline proposals

Section A
Cost planning during feasibility (page 157)

Section B
Cost planning during outline proposals (page 187)

Section C
Cost planning during scheme design (page 202)

To be effective, cost planning should be performed in steps of increasing detail which correspond with the increasing and associated information detail of the design.

If major cost differences between the buildings analysed and the new project are not assessed as soon as they become apparent, 'cost control' will not be as effective as it could be. (In general, the cost analyses chosen at feasibility are also used for cost planning at outline proposals and scheme design.)

What information is usually available at the beginning of outline proposals which would not usually be available at the beginning of the feasibility stage?

Design criteria. *Turn to page 194*

Outline design. *Turn to page 191*

Both of these. *Turn to page 195*

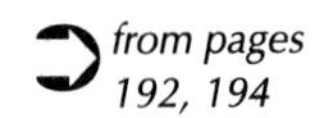
from pages 192, 194

You are incorrect.

This would be a waste of time. It might indicate whether or not the cost limit would be exceeded when more reliable cost targets were prepared during scheme design, but it would be impractical to take remedial action on the basis of cost targets which nobody was taking very seriously.

Remember, cost decisions should be as detailed as the available information allows, and no more. The objective at outline proposals is to allocate the available money to the various parts of the proposed building, without making any design assumptions which may be impossible to maintain at a later stage.

Reconsider your answer to the question: how is an outline cost plan prepared?

Choose one of the following answers.

Cost targets are prepared for each major group of elements. *Turn to page 197*

Allowances are made for each of the major differences, and tabulated. *Turn to page 192*

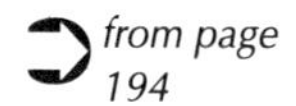
from page 194

Design criteria for office block

Site

(1) *Site information and ground conditions*

(a) Levels. Site plans would normally be available at this stage indicating relevant site features and including a grid of levels or the site contours.
(b) Soil. Clay and gravel, site investigation has been conducted.
(c) Bearing capacity. 200 kN/m^2.
(d) Water table. None encountered.
(e) Depth of bearing strata. For three-storey block – 1.25 metres.
(f) General. Site cleared of old terrace houses without basements. Foundations grubbed up under a separate demolition contract.
(g) Utilities. Main drainage available, water main in road adjoining the site. Electricity supply available. No undergound utilities likely to cause difficulty.

General requirements

(2) *The building*
The building has to conform to the provisions of the Building Regulations. In particular, the structure and protected shafts shall be of 1.5 hours fire resistance. Doors to protected shafts shall be of 1 hour fire resistance.

(3) *Working conditions*
Temperature, ventilation, lighting, sanitary conveniences, etc. shall comply with the latest requirements of current legislation.

Turn to page 190

from page 189

Structure

(4) *Frame*

Storey height for all floors	3.5 m
Span	approximately 15 m
Fire resistance	1 hour

(5) *Upper floors*

Imposed load	5 kN/m^2
Fire resistance	1 hour

(6) *Roof*

Imposed load	0.75 kN/m^2

(7) *Internal doors*

Doors to protected shafts	1 hour fire resistance

Environmental engineering services

(8) *Heating and ventilation*

Air-conditioning using four-pipe fan coil comfort cooling system.

No. of air changes per hour	3
Temperature	22°C (dry bulb) 50% RH

(9) *Electrics*

Lighting standard	500 lux

(10) *Other services*

All normal services to typical commercial standards including:

- ❑ hot and cold water supply
- ❑ fire services
- ❑ small power
- ❑ provision for communications
- ❑ lightning protection
- ❑ lifts
- ❑ drainage, etc.

Return to page 194

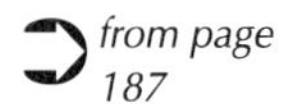
from page 187

You are incorrect. The outline design is prepared *during* the outline proposals stage.

Are the design criteria usually available at the beginning of outline proposals?

Write down your answer.

Turn to page 194

from pages 188, 194

You are incorrect.

Major differences are certainly isolated and allowed for. However, this only allows for confirmation or amendment of the cost limit.

The objective of cost planning in outline proposals is usually to start allocating the available money to the various parts of the proposed building.

Reconsider your answer to the question: how is an outline cost plan prepared? Choose one of the following answers.

Cost targets are prepared for each element on the assumption that the inaccuracies in the unreliable ones will cancel each other out.

Turn to page 188

Cost targets are prepared for each major group of elements.

Turn to page 197

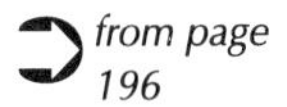
from page 196

Each target cost is established by interpolation from the cost and information contained in the cost analyses (bearing in mind the requirements of the new project). Major differences, between the new project and the cost analyses must be allowed for and included within the appropriate group element cost targets.

The build-up of the outline cost plan is carried out systematically, starting with the substructure and finishing with the preliminaries. An allowance needs to be included for contingencies in order to accommodate costs associated with unforeseen events or circumstances which may or may not occur during the construction stage of the project.

Finally the price and design risk is reworked and added to the outline cost plan.

The total of the new project is arrived at by multiplying the cost per m^2 of gross floor area by the gross floor area of the new project.

During the preparation of the outline cost plan, the various allowances should obviously be made with an eye on the cost limit.

Minor juggling of figures and reworking of calculations may be necessary to arrive at the final figure to be reported to the client.

When the outline cost plan has been completed, it is sent with the outline design for the client's approval. Obviously the outline cost plan is completed by the end of the outline proposals stage.

When is the outline cost plan started: before, during, or after the preparation of the outline design?

Write down your answer.

Turn to page 198

from pages 187, 190, 191, 195

Yes. Design criteria are usually available at the beginning of outline proposals.

If you would like to see design criteria for the worked example in Chapter 11, turn to page 189.

The other information which is usually available at the beginning of outline proposals is:

(1) the general brief;
(2) the local authority planning policy;
(3) the feasibility report (including the first estimate, which has been approved by the client by this time).

The increased amount of information makes it possible to isolate several differences between the buildings analysed and the new project which cannot be spotted at feasibility. However, there is still not enough data for reliable cost targets to be prepared for each element. Instead, an 'outline cost plan' is prepared at this stage.

Write a brief description of how an outline cost plan is prepared, then choose the answer which corresponds most closely with yours.

(a) **Cost targets are prepared for each element on the assumption that the inaccuracies in the unreliable ones will cancel each other out.** *Turn to page 188*

(b) **Cost targets are prepared for each major group of elements.** *Turn to page 197*

(c) **Allowances are made for each of the major differences, and tabulated.** *Turn to page 192*

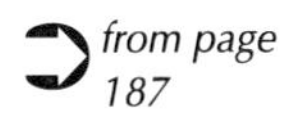
from page 187

You are incorrect. The outline design is prepared *during* the outline proposals stage.

Are the design criteria usually available at the beginning of outline proposals?

Write down your answer.

Turn to page 194

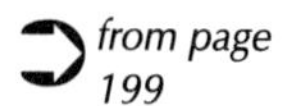
from page 199

The first step is to gather together again the 4 to 10 cost analyses used for the feasibility stage. It is possible that new data will have become available during the interval but more often than not the data source does not change.

The adjustments for time must be carried out again to update all of the cost analyses to the current date. The current index used at feasibility stage may no longer be appropriate.

The next step in preparing the cost target for a group of elements is to again 'isolate the differences'. This is done by comparing the specifications of the elements in the analyses with what is known about elements in the new project.

These differences can be allowed for by the methods already described when making adjustments during the preparation of the first estimate or by methods used for the preparation of cost targets for individual elements. This will be illustrated in Chapter 11. All that need concern us here is that the allowances are more detailed than those made during the feasibility stage and less detailed than those made during scheme design. The reason for this, of course, is that the information on which allowances are based becomes increasingly detailed as the design process proceeds.

The quantities of the various elements are not usually known at this stage. Differences in the sizes of the buildings are usually allowed for by considering the cost of the group of elements per m^2 of gross floor area.

How would the cost targets for each group be obtained from this information?

Write down your answer.

Turn to page 193

from pages 188, 192, 194

You are correct. Cost targets are prepared for each major group of elements.

There is an important distinction between an outline cost plan and a tabulation of allowances for the major differences which become apparent at outline proposals. With the outline cost plan, a cost target is prepared for each major group of elements. (The 'groups of elements' considered in the worked example in Chapter 11 are those identified in cost analyses i.e. substructures, superstructures, finishes, fittings, services, external works, preliminaries and contingencies.) This has the advantage that it forces the design team to think clearly at this early stage about the broad allocation of expenditure to the major parts of the building.

Before preparing an outline cost plan, the reserve against price and design risk should be considered again. This is the allowance made for unforeseen design difficulties which may only come to light later in the design process and price rises occurring between the preparation of the outline cost plan and the receipt of tenders.

If a cost limit has been set at the feasibility stage, it is advisable to reconsider the reserve against the price and design risk; this should be carried out before group cost targets are prepared. This is done to identify clearly the sum that has been estimated at the feasibility stage as appropriate for distribution among the group elements, thus providing guidance when calculating cost targets for each group element. Any minor inaccuracies and discrepancies between the outline cost plan and the cost limit are then usually accounted for by recalculating and adjusting the price and design risk allowance.

Turn to page 199

from page 193

Ideally designing and costing should be done at the same time. Moreover, there should be considerable interplay between the two, as indicated on page 153, in order to maintain strict control over costs (i.e. with every cost decision being allowed to affect the corresponding design decision, and vice versa).

We realise that this is an ideal approach which might not happen in practice, but it is always better to aim higher than one really expects to achieve.

For the purposes of demonstration in Chapter 11, it has been necessary to assume that the outline design has been completed before the outline cost plan is started. We have done this only to eliminate complicated descriptions of the timing and the complex decision-making of a properly conducted outline proposals stage.

Briefly describe the cost allocations and allowances that are included in an outline cost plan, and give brief details of how they are prepared.

(Refer back to pages 196 and 193 if necessary.)

Turn to page 200

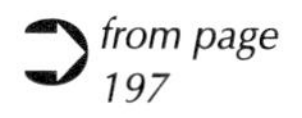
from page 197

Because time has moved on, the allowance now made for price rises may well differ from the corresponding allowance included in the feasibility estimate. If you would like to see the 'price and design risk' allowance from the worked example in Chapter 11, turn to page 242.

We will now consider how the cost targets are prepared for each group of elements. (The 'groups of elements' considered in the worked example in Chapter 11 are those identified in cost analyses i.e. substructures, superstructures, finishes, fittings, services, external works, preliminaries and contingencies.) When discussing cost planning during the feasibility stage, we stated that research has indicated that it is a mistake to focus on one cost analysis; we suggested using between 4 and 10 analysed projects. Those same analyses should, wherever possible, be used to prepare cost targets for each group of elements.

What will be the first step in the preparation of a cost target for a group of elements?

Write down your answer.

Turn to page 196

from page 198

Summary – Cost planning during outline proposals

(1) An allowance is made for 'price and design risk' by assessing the cost of unforeseen design difficulties and possible price rises between the preparation of the outline cost plan and the receipt of tender.

(2) Cost targets are prepared for each group of elements. Major differences between the buildings analysed and the new project must be estimated and included in the appropriate group of element target costs. The group elements must include a target cost for preliminaries.
The major differences are allowed for by comparing all the information available for both the new project and the cost analyses.

(3) An allowance is made for contingencies based on assessment of the risk of unforeseen events or circumstances which might arise during the construction stage. This specifically recognises that the construction stage has specific risks. For example, even though the site investigation has shown the expected ground conditions, there may be 'soft spots' or rock encountered on part of the site which will require more money to be spent.

Turn to page 201

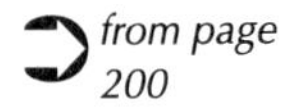
from page 200

A ***major discrepancy*** may occur between the outline cost plan and the cost limit. Here, it is best for the ***client and the design team to make a* joint *decision*** as to whether it is better to juggle with the allowances or to accept the outline cost plan as a more realistic total. If this results in the cost limit being raised rather than lowered, at least the client has played an active part in the decision and has been afforded the opportunity of abandoning the project before incurring major expenditure.

When the client accepts the outline cost plan, the design team is committed to producing a scheme design within the cost shown on the outline cost plan.

We shall now discuss the cost planning techniques used during the scheme design stage.

Turn to page 202

from page 201

Section C

Cost planning during scheme design

Section A
Cost planning during feasibility (page 157)

Section B
Cost planning during outline proposals (page 187)

Section C
Cost planning during scheme design (page 202)

Turn to page 203

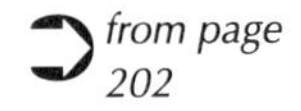
from page 202

It is best to start preparing the detailed cost plan as early as possible during this stage. Cost targets prepared early on can be checked after the scheme design drawings have been completed. The main data source available to help in preparing the cost targets will be the same 4 to 10 cost analyses used previously. As far as details of the new project are concerned, the information now available includes:

(1) the complete brief;
(2) the outline design (including drawings);
(3) the outline cost plan;
(4) all the information available at the beginning of outline proposals.

The cost plan is finalised at the end of the scheme design stage, by which time all of the scheme design drawings will also be available.

Before considering each element individually, the reserve against price and design risk is reconsidered. As at the outline proposals stage, the object is to give guidance. Subtracting this reserve from the cost shown on the outline cost plan leaves the balance available for cost targets for individual elements. Once cost targets have been set, the price and design risk reserve will be recalculated, bearing in mind the cost limit.

In general, would you expect the allowance for 'price and design risk' at scheme design to be the same size as the one made during outline proposals?

Write down your answer.

Turn to page 207

from page 205

Differences in quantity and quality will vary from element to element. Differences in price level are more likely to affect all elements equally. Therefore, it is worth considering the allowances made for price level before we go any further.

With a computer system, price level adjustments can be handled speedily and almost automatically. Often, the only information required is the tender date, the location of the buildings analysed, and the current date and location of the proposed project. It is important, however, to understand what is involved and thus further consideration will now be given to price level adjustments.

The factors listed under 'price level' in Chapter 9 were:

(a) differences in general tender price level due to difference in tender dates;
(b) differences in contractors' price levels relative to the general tender price level (due to differences in allowances for overheads, profits, etc.);
(c) differences in regional trends, and localised variables;
(d) differences in tendering arrangements (including selection of contractors for tendering lists), contract conditions, contract period, weather conditions, site conditions, etc.

Factor (a) is the most important and probably the easiest to adjust.

Consider the changes in general tender price level between the date of tender of the buildings analysed and the proposed ***receipt of tender of the new project***.

Can these be allowed for accurately by using the relevant values of a tender price level index?

Yes/No.

Write down your answer.

Turn to page 208

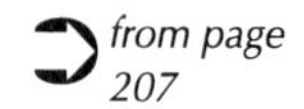
from page 207

The three factors affecting element costs are:

Factor 1 Quantity.
Factor 2 Quality.
Factor 3 Price level.

Each elemental cost target should be prepared by isolating differences in each factor between the project analysed and the new project.

The task is similar to that carried out at outline proposals stage. However, there is more information to handle as each element, rather than a group of elements, is considered.

Not all of the selected cost analyses will contain similar or suitable design details. Thus some cost items will be inappropriate when considered on an element-by-element basis, which is not a problem if the data source is wide enough to allow a selective choice to be made.

Computer systems greatly aid the process when a lot of information needs to be investigated and manipulated.

Turn to page 204

from page 208

(1) **Past to present. The differences between the general tender price levels at the dates of tender of the buildings analysed and at the date of preparation of the cost plan are allowed for by using an appropriate index to update each element in turn.**
(2) **Present to future. The difference between the general tender price level at the date of preparation of the cost plan and the contractor's price level at the proposed tender date is allowed for in the 'price risk'.**

A diagram illustrates this:

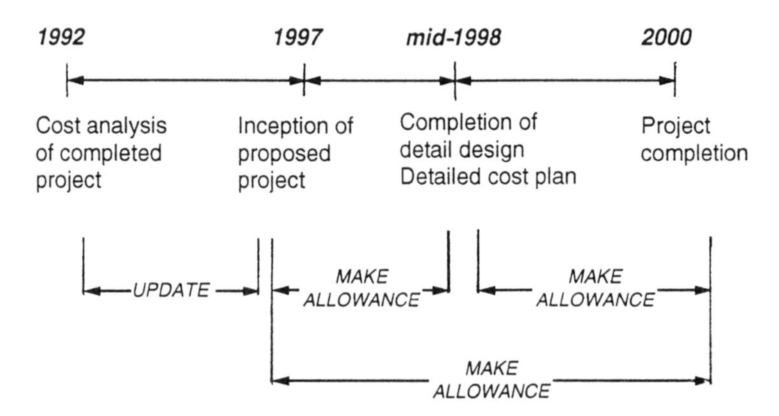

The difference in general tender price level is not the only 'price level' factor which is considered during the preparation of individual cost targets. As well as ***updating*** the price levels of the buildings analysed, these prices are also adjusted for ***location***, using a location index.

Any of the other significant 'price level' factors should be considered and adjusted at this stage.

Obviously all the factors affecting element costs should be considered when an elemental cost target is being prepared for each element. Write down these factors.

Turn to page 209

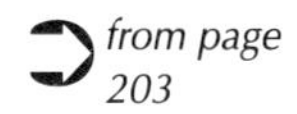
from page 203

The price and design risk reserve at scheme design will usually be slightly less than at outline proposals. The design has advanced considerably since the preparation of the outline cost plan. Therefore, the 'design risk' part of the allowance can be decreased safely. Time will also have moved on and the index previously used should be checked to see if it still holds good. Any movement will be reflected in that part of the reserve dealing with price risk.

The order in which the elements are considered will vary from project to project. One general rule is that preliminaries should be considered before any of the other elements. One reason for this is that preliminaries and contingencies can be spread over other elements instead of being considered as separate elements. The design team should decide as early as possible which method they intend adopting, as the outcome will have an appreciable effect on the level of pricing of all the other elements. The commonest practice, and the approach we adopt, is to consider preliminaries as a separate element.

Another reason for considering preliminaries first is that the contractor's profits on the other elements are sometimes included in preliminaries. It is obviously necessary to find out if this has been done in the cost analyses before considering the other elements. It is also more convenient to deal with contingencies at this time.

The order in which other elements are considered will usually be dictated by the order in which the scheme design drawings are produced. In Chapter 11 we have assumed that the scheme design drawings are available before the cost plan is started. Therefore, we have considered the elements in numerical order, apart from preliminaries and contingencies.

Turn to page 205

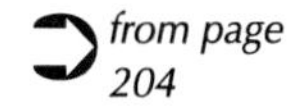
from page 204

No.

The general tender price level index at the date of tender of the new project is only a forecast. Updating on the basis of this index may not be necessarily, so accurate.

Instead of relying on this future value of the index, we use the index at the date of preparation of the cost plan. This means that the cost targets incorporate realistic allowances for general tender price level changes up to the date of preparation of the cost plan.

The change in the general tender price level between the date of preparation of the cost plan and receipt of tender is allowed for in a lump sum. This includes an allowance for possible differences between the contractor's price level and the general tender price level at the date of tender.

This lump sum is incorporated in the 'price risk' allowance.

Thus, the difference between the general tender price levels for the buildings analysed and the contractor's actual price level for the new project is allowed for in two stages.

Write down brief descriptions of these two stages.

Turn to page 206

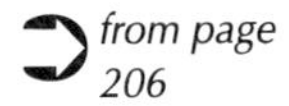
from page 206

There are three factors affecting element costs. These are allowed for in each of the methods used to prepare cost targets.

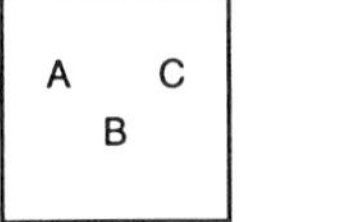

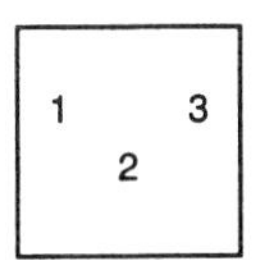

Factors
A Quantity
B Quality
C Price level

Methods
1 Simple proportion
2 Inspection
3 Approximate quantities

Simple proportion is preferred to the other two methods because a simple proportionate adjustment to an element cost automatically includes allowances for everything needed for the element in the type of building being considered.

When using approximate quantities it is easy to overlook details which will only be considered during the detail design stage. Nevertheless, this method sometimes has to be used, especially when a new requirement is under consideration.

Inspection is usually reserved for elements for which no effective measure of Quantity has been devised. These include fittings and furnishings or builder's work in connection with engineering services. Often the elements are not cost significant. In general, this should not be a criterion, as simple proportion takes very little time to complete provided there are satisfactory measures of element unit quantity.

Each of these techniques is now considered in detail, starting with 'simple proportion'.

Turn to page 210

from page 209

Simple proportion

This technique is called 'simple proportion' because it assumes that the cost of an element is directly proportional to the Quantity of the element.

Assuming that the cost of an element is directly proportional to its Quantity, then we can say:

1 Simple proportion
2 Inspection
3 Approximate quantities

$$\frac{\text{cost of element in new project}}{\text{quantity of element in new project}} = \frac{\text{cost of element in building analysed}}{\text{quantity of element in building analysed}}$$

We can now use this expression to calculate the cost of the element in the new building.

Example

Building analysed:	gross floor area 5862 m^2
Element:	internal walls and partitions
Total cost of element:	£323 270
Element unit quantity:	5142 m^2
Proposed project:	gross floor area 6000 m^2
Element:	internal walls and partitions
Element unit quantity:	5500 m^2

$$\text{Cost of element in new project} = \frac{£323\,270}{5142\,\text{m}^2} \times 5500\,\text{m}^2$$
$$= £345\,777$$

This is the same as saying that:

element unit rate in new project = element unit rate in analysis

However, we are getting ahead of ourselves because we said that there were three factors that affect cost targets.

What are the three factors that affect cost targets?

Write down your answer.

Turn to page 212

from page 212

2 *Adjust for Quantity*

This section started by discussing the adjustment for Quantity. We noted that it is done simply by multiplying the appropriate element unit rate by the Quantity of the element in the proposed project.

This figure is then reduced to cost per m^2 of gross floor area.

Thus, returning to the example on page 210, we have:

$$\frac{£345\,777}{6000\,m^2} = £57.63 \text{ per } m^2 \text{ of gross floor area}$$

The simple proportion method should be the commonest approach employed but its use is not unlimited.

Write brief descriptions of two circumstances when the simple proportion method cannot be used for preparing a cost target.

Turn to page 213

from page 210

The preparation of a cost target takes place in three stages to account for the three factors affecting costs:

Factor 1 Adjust for Price level.
Factor 2 Adjust for Quantity.
Factor 3 Adjust for Quality.

(We have reversed the order here to follow common practice, which is to adjust for price level first.)

1 *Adjust for Price level*

The differing tender dates and various locations of projects analysed will usually mean that 'adjusting for price' is essential. The assumption is that the method of simple proportion is not invalidated by these adjustments. Consider first adjusting for differing tender dates.

In terms of simple proportion this can be stated as:

$$\frac{\text{current element cost}}{\text{current index}} = \frac{\text{element cost in analysis}}{\text{index at analysis tender date}}$$

In more conventional terms, the adjustment required can be shown as:

Element cost in analysis  'Updating' index  'Location' index

The adjustments for price and location are reasonably straightforward and should be ***completed first***. This ensures that current prices applicable to the project location are used at each stage of the preparation of a target cost.

Turn to page 211

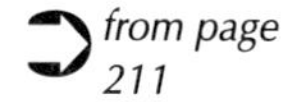
from page 211

Simple proportion cannot be used when

(a) **there is no effective measure of Quantity for the element (i.e. no element unit quantity/rate exists), or**
(b) **it is not yet known what Quantity of the element will be required in the new project.**

In these cases the possibility of using 'inspection' and/or 'approximate quantities' would be explored. (Both are explained later.)

3 *Adjust for Quality*

This is usually the least straightforward of the three adjustments. During scheme design the quality of each element cannot be specified with any certainty. It is not until the detail design stage that the precise specification is known. At the scheme design stage it is necessary to ensure that the cost target for each element is neither too generous nor too small for the *general* quality of the materials which the design team intends to adopt. It might seem rather difficult to determine the cost of the 'general quality' before action is given to consider the details of specification of each element. When the simple proportion method is used, however, the design team do not have to start from scratch when they try to fix a cost to their vague notions of the quality which they intend to specify in the detail design stage.

Write a brief description of why this is so.

Turn to page 215

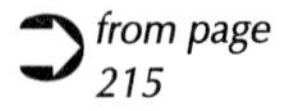
from page 215

Rather than decide by inspection on a percentage increase or decrease to allow for a change in quality, approximate quantities are often used. That is, the detailed design of an analysed building is altered for an element in the cost analysis which does not seem to be of a suitable quality for the new project. Approximate quantities are measured from this design, and priced in the traditional way. The change in cost is then expressed as a percentage of the total element cost in the analysis. This percentage is added to (or deducted from) the figure obtained by adjusting for Price and Quantity.

(Note: This method does not commit the design team to using this specification for the part of the element concerned, but simply to selecting one of a similar general quality.)

You may have noticed that variations in the quantities of individual items *within* each element under 'adjust for Quantity' have not been discussed. This is considered under adjustment for quality and is best explained by an example.

Example

Changing the specification of the floor finishes of a building from 10% granolithic and 90% wood block to 90% granolithic and 10% wood block will obviously change the quality of an element. However, this will not necessarily affect the element unit quantity of the floor finishes.

This is exactly the kind of change in specification which could be handled by the 'inspection' or 'approximate quantities' methods.

Throughout this programme the term 'quality of an element' is used to mean both the quality of the materials and the relative proportions of the various qualities of specification within the element.

Turn to page 216

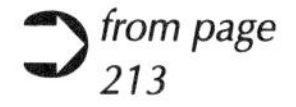
from page 213

Adjusting the cost of an element for Price and Quantity gives the current cost if exactly the same detailed specification is used. By examining the specification in the cost analyses, the design team can usually spot any elements which differ in quality from those they envisage using in the new project. That is, it is easier to spot differences and make realistic allowances for them, than it would be to fix a cost to partially-formed ideas about the new project on their own.

The adjustments made for a change in the quality of an element are often expressed as percentage additions to (or deductions from) the costs obtained after adjusting for Price and Quantity.

Example: Element 2D Stairs
The building analysed has steel tubular balustrades with PVC sheathed handrails. The design team decide that they would like to have a better quality of handrail and balustrade in the new project. Bearing in mind their cost limit and the requirements of the other elements, the design team decide to increase the allowance for the element by 15%. During the detail design stage the exact type of balustrade will be determined.

The figure obtained from the cost analysis after adjusting for Price and Quantity is: £54 366.

The cost target is therefore:

£54 366 + 15% = £62 521

Turn to page 214

from page 214

While preparing a cost plan, there is a need to ensure that the cost targets are reasonable. The obvious way to do this is to compare the cost targets with the costs of corresponding elements within other analyses. Comparing total costs, such as £31 837 and £45 081 for upper floors, is not very informative. The costs per m^2 of floor area or the element unit rates are more meaningful and so they will be used for this purpose.

However, bear in mind that costs per m^2 of gross floor area reflect differences in density, i.e.

$$\frac{\text{element unit quantity}}{\text{gross floor area}}$$

as well as Quality and Price level, whereas element unit rates do not reflect all these differences. (See Chapter 9 page 126 for a discussion on this point.)

For this reason, it is recommended that both the total target costs and the target costs per m^2 of gross floor area should be entered in the cost plan summary. This approach has the added advantage of expressing costs in a way which allows comparisons of costs both within one building and between buildings of different sizes.

Before moving on to the next method of adjustment (inspection), complete the following:

Write down the names and a brief description of the three adjustments made during the preparation of a cost target by simple proportions.

Turn to page 217

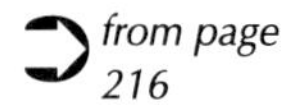
from page 216

Summary – Adjustments made during simple proportion method

The adjustment is done in three stages:

Stage 1 — ***Adjust for Price*. The selected element unit rate from the cost analysis is updated and adjusted for location.**

Stage 2 — ***Adjust for Quantity*. The adjusted element unit rate is then multiplied by the Quantity of the element for the proposed project. This is reduced to a cost per m^2 of gross floor area by dividing it by the gross floor area of the proposed building.**

Stage 3 — ***Adjust for Quality*. Differences in either the quality of materials or the relative proportions of the various qualities of specification within the element are allowed for by adjusting the target cost and cost per m^2 gross floor area obtained above.**

The total target cost and the target cost per m^2 of gross floor area are entered in the cost plan summary.

(*Note:* The Price and Quantity adjustments are usually done before the Quality adjustment because:

(a) they always have to be made, and tend to have more effect on the figure than the Quality adjustment,

(b) the quality of the element is easier to change later in the design process than the Quantity of the element.)

Turn to page 218

from page 217

2 Inspection

This consists of assembling a range of costs (£/m^2 gfa) of corresponding elements in previous projects. From this a suitable cost target can be chosen.

(If you would like to see an example of this technique, turn to page 219.)

1 Simple proportion
2 Inspection
3 Approximate quantities

When is 'inspection' used? Choose the best criterion from those listed below.

(a) When simple proportion cannot be used (as discussed on page 213) and the cost involved does not justify the time required for measuring approximate quantities.
(b) When simple proportion cannot be used and the final specification of the element is likely to be similar to the specifications of the element in other projects whose cost analyses are available.
(c) When measures of Quantity are not available and considerable time would be required for measurement of approximate quantities, and the quality standard of the element in question is not unusual.
(d) When the cost involved does not justify the time required for either simple proportion or approximate quantities.

Write down your answer: (a), (b), (c), or (d).

Turn to page 220

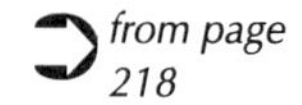
from page 218

Example from Chapter 11 using 'inspection'

5C Disposal installations

This is not a major element in this project which is best decided by inspection. (Approximate quantities could also be used if the cost involved justified the time required for measurement.)

The engineering services in our project are located in close proximity to staircase areas. Thus reasonably short horizontal pipe runs can be expected. We need to remember that storey heights reflect the incorporation of suspended ceilings and raised floors in our design. These points need to be considered when inspecting similar cost analyses and other data sources.

The costs of disposal installations in existing offices are as follows:

Building	Tender date	Prelims	Cost of disposal installations (£/m^2 gfa)
Two-storey offices, *Winchester*	Mar. 1991	10.27%	3.50
Four-storey offices, *Manchester*	Feb. 1993	7.66%	4.86
Two-storey offices, *Slough*	Jan. 1993	8.79%	4.01
Two-storey offices, *Reading*	Sept. 1992	9.50%	3.89

Making approximate adjustments for the differences in the preliminaries percentages and the tender dates of these projects, we decide to adopt an allowance of £4.75 per m^2 for the element which includes builder's work to pipe ducts.

Total element allowance = £4.75 x 2246 = £10 669

	Total	**Cost per m^2 gfa**
Cost target for disposal installations	**£10 669**	**£4.75**

Return to page 218

from page 218

Both (a) and (b) are relevant.

- **(c) is the best criterion as it comprises (a) and (b).**
- **(b) is relevant because we prefer to use only approximate quantities when an element in the proposed building is going to be unlike its counterpart in any other buildings whose cost analyses are available.**
- **(d) is a non-starter: simple proportion takes very little time and should always be used when satisfactory measures of element unit quantity are available.**

We now go on to consider the technique of 'approximate quantities'.

Turn to page 221

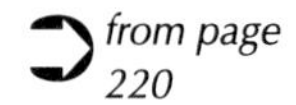
from page 220

Approximate quantities

1 Simple proportion
2 Inspections
3 Approximate quantities

Under the method of 'approximate quantities' a relatively detailed specification is prepared for the element in question. Approximate quantities are then measured from this specification and priced in the traditional way.

(If you would like to see an example of this technique, turn to page 222.)

It should be noted that:

(a) the specification prepared for this technique is not as detailed as that prepared during the detail design stage,
(b) when using approximate quantities it is easy to overlook details which will only be considered during the detail design stage,
(c) only the major or most cost-significant items in any element are measured and priced; the costs associated with sundry items must be included with the rates or alternatively they may be included by adding an overall percentage at the end,
(d) the design team is not committed to finally using the specification prepared for approximate quantities, but simply to using one of a similar general quality.

Write a brief description of the one situation in which approximate quantities should, theoretically, be used in preference to the other two techniques.

Turn to page 223

from page 221

Example from Chapter 11 using 'approximate quantities'

2G: Internal walls and partitions

The building analysed had demountable partitions. Most of the partitioning is left as a tenant's responsibility in our project. In addition, our standards are somewhat different and we also have the effect of increased storey heights to take into account. We therefore decide to use the approximate quantities technique.

73 m^2 of 100 mm blockwork partition @ £21.00 per m^2	£1533
255 m^2 of 200 mm blockwork partition @ £32.00 per m^2	£8160
77 m^2 of 112.5 mm common brick internal walls @ £23.00 per m^2	£1771
18 WC cubicles @ £330 each	£5940
	£17 404

Reduce to cost per m^2 of gross floor area: $\frac{£17\,404}{2246\,m^2} = £7.75/m^2$

	Total	Cost per m^2 gfa
Cost target for internal walls and partitions	**£17 404**	**£7.75**

Return to page 221

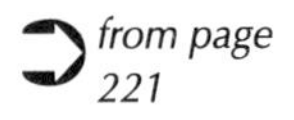
from page 221

Theoretically, approximate quantities should only be used when an element in the proposed building is markedly different from its counterpart in all the other building analyses.

However, in practice approximate quantities can be used for all elements without destroying the usefulness of the cost plan. Using approximate quantities throughout would not violate the principles of cost control. Nevertheless, the cost plan will be prepared more quickly and cheaply, and may well be more realistic if the other two techniques are used.

Write down the names and brief descriptions of the three techniques which can be used for deciding on the cost target for an element.

Turn to page 224

from page 223

Summary – Methods of preparing cost targets for elements

1 *Simple proportion*

The element cost selected from a range of cost analyses chosen at feasibility is adjusted by proportion.

2 *Inspection*

A target cost per m^2 of gross floor area is chosen from a cost range obtained from a selection of suitable analyses and/or cost studies.

3 *Approximate quantities*

A provisional specification of the required quality is prepared for the element and priced by approximate quantities.

(*Note:* No matter which technique is used, the resulting cost target should be regarded as a reasonable allowance within which the element may be designed during the detail design stage. The design team should never feel obliged to use exactly the same specification upon which the cost target is based.)

Turn to page 225

from page 224

Each of these methods of preparing cost targets takes into account all the factors which affect element costs.

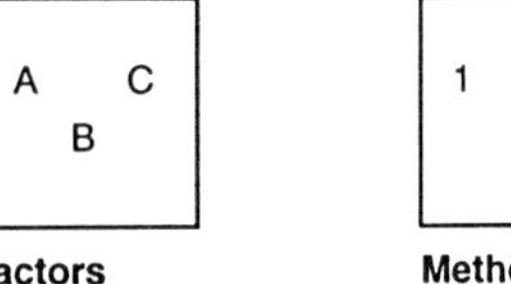

When all the cost targets have been entered in the cost plan summary, their total must be compared with the cost limit and remedial action should be taken if necessary.

It might seem that the preparation of a cost plan takes rather a long time. In practice, it can take less time than you have spent reading this chapter!

Chapter 11 is devoted entirely to an example of cost planning from feasibility to scheme design. This has been inserted before Chapter 12 to give you a fuller appreciation of the 'planning' part of cost control before going on to consider the control of costs during the detail design stage.

Turn to page 227

11 Cost Example from Feasibility to Scheme Design

This chapter *is not* in programmed form.

This chapter is a step-by-step example of how the cost plan develops as more design information becomes available. It starts with the first estimate at the feasibility stage which we have called FS at the top of the page to help guide you through the chapter.

It moves on to the outline cost plan at the outline proposal stage – marked OP at the top of the page.

The last example is the detailed cost plan at the scheme design stage (marked SD at the top of the page). Remember the detailed cost plan is finalised at the end of the scheme design.

The example in this chapter illustrates the cost planning of an office block from the preparation of the first estimate at the feasibility stage to the detailed cost plan at the scheme design stage.

In order to keep this exercise within manageable limits we have selected just one analysis rather than a range of analyses as recommended.

The example is presented in the form of a running commentary by the design team during each of the three design stages involved. When you read a phrase like 'We decide to allow £1000', imagine yourself swept back in time to the date quoted at the beginning of that particular stage of the design.

The relevant dates are:

First estimate	June 1995
Outline cost plan	September 1995
Detailed cost plan	November 1995
Proposed tender date	September 1996

FS

Feasibility stage (June 1995)

The information available at this stage is normally very limited and, in this instance, includes:

(1) Initial brief identifying the use to which the building is to be put, the floor area required, and an indication of the desired quality.
(2) Town planning policy.
(3) Details of the site.

In addition the following data will be used:

- The histogram of average building prices; new office buildings generally.
- The detailed cost analysis of the office block, given in Appendix A.
- Access to indices, for adjusting price level (not included here).

The above information and data has been assembled before we start to prepare our Feasibility stage estimate.

Initial brief for proposed office premises*

Space	A lettable area of at least 1850 m^2 and maximum permissible car parking. The parking provision is considered to be essential due to the congested city centre site, poor parking facilities and inadequate public transport. The tenants are likely to be professional services such as accounting and insurances.
Planning permission	Preliminary enquiries to the planning authority indicate that an office block three storeys high would be permitted and that the stated amount of lettable area should be achievable as it is within the plot ratio.
Use	The premises will be offered for leasing as good-quality office accommodation at a rental which is in the upper market range.
Quality	An above-average commercial quality is appropriate to this prime location site and will command a suitable return. Raised floors and suspended ceilings will need to be incorporated into the new design in order to accommodate IT systems. An attractive façade is considered essential.
Site	The site has been acquired and a site plan is available. (It has not been included here as it is reasonably straightforward.) It shows that the terrace houses previously occupying the site have been cleared and the foundations grubbed up. Ground conditions in the area are good.

* Remember, this is the ***only*** information available at this stage

FS

Estimate

We must prepare our feasibility stage estimate so that the client can decide whether or not to go ahead with the project. This follows examination of the client's initial brief, an inspection of the site and local conditions, and agreement of quality standards between the design team and the client.

It is possible to gain an idea of the cost range by examining analyses of past projects for the same type of building. This information may be obtained from in-house records, or from published information. An information service, such as BCIS, publishes a wide range of suitable cost information. The histogram in Fig. 11.1 shows the cost range for office buildings, which has been updated to the 4th quarter 1995. The costs are for the building works and exclude external works, drainage, external services, professional fees and VAT. The mode is £550–£600 per m^2 gfa.

Experience and knowledge is needed to interpret these cost ranges. For example, higher quality finishes and fittings to the ground floor reception, or enhanced finishes and services for restaurants and computer suites may all increase the overall cost per m^2 gfa. Similarly, allowance must be made for the cost of any demolition and enabling works and the costs of site investigation, planning and Building Regulations approvals and work to adjoining properties.

Remember, costs will vary according to location, quality, and size of development.

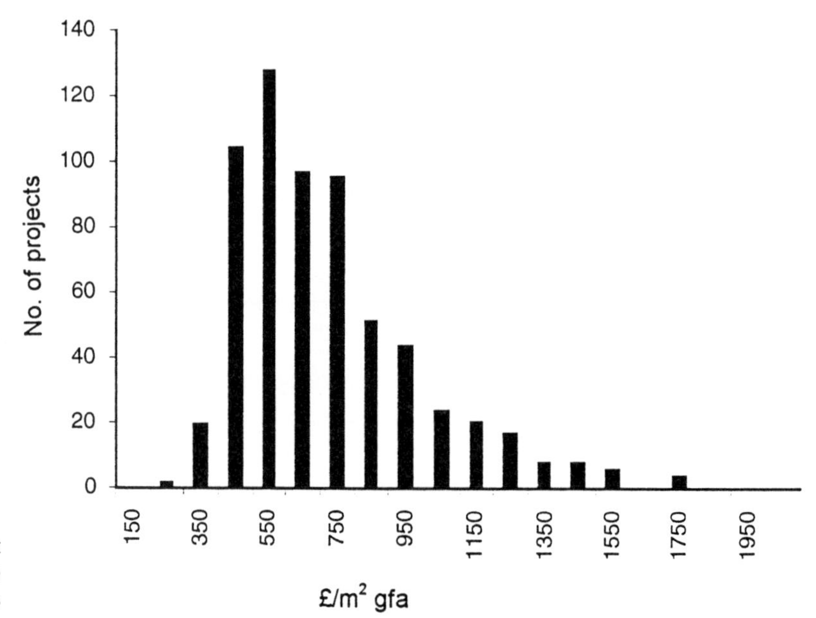

Fig. 11.1 Cost range for office buildings

FS

From this distribution, it appears that we should expect our estimate to be in the range £450 to £850 per m^2, which is fairly wide. This range is influenced by many factors including the exact quality desired (note that these prices do not include any allowance for external works). Since this histogram tells us nothing about the specific characteristics or quality of the buildings concerned, we must treat this type of study with caution. In fact some m^2 prices have been recorded well above £850.

It is, therefore, preferable to work from detailed cost analyses which provide some information about the buildings analysed. Ideally a small number (between 4 and 10) of similar analyses should be selected. This will permit adjustments to be made for items where our brief or quality standards depart from those of the cost analyses. We should obviously select analyses which are reasonably close to both our functional requirements and our standards of quality.

To keep this exercise within manageable limits we have selected just one analysis (given in Appendix A) which is in our opinion suitable with regard to architectural style and quality. Ideally, in order to provide a good match for the new project, this cost analysis should have been a three-storey rather than a two-storey construction. Also the storey heights of the chosen analysis appear to be too low for our new project, nevertheless, this does allow us to consider some important design variables and their implications for project costs. The analysis has, therefore, been accepted for our purposes.

Before considering the major differences between the building analysed and our project, we need to estimate the total floor area which will be required for our project. From studies of other office blocks, it appears reasonable to allow 20% of the lettable floor area for the additional area required for services and circulation space.

Lettable floor area required		$= 1850\,m^2$
Add		
Allowance for circulation space (i.e. corridors, toilets, stairwells, lift wells, etc.),		
$1850 \times 20\%$		$370\,m^2$
		$2220\,m^2$
	say	$2250\,m^2$

We will assume a gross floor area of $2250\,m^2$ for our project. Until the preparation of scheme design drawings, what is required now is a suitable all-inclusive rate to calculate our feasibility estimate. From our chosen analysis, the cost per m^2 gfa (at September 1992) for the building sub-total is £612.22 per m^2. Total cost (less contingencies) is £819.46 per m^2. We cannot use these figures without first making some necessary adjustments for Price level, Quantity and Quality.

FS

It is important to remember that at this stage we have very little information either in the form of specification and design notes or as drawn information. Bearing this in mind we shall look at the adjustments.

Adjustment for Price level

This is the adjustment for the factors listed in Chapters 9 and 10. Of these, time and location are the most important.

Adjustment for time:

Index at the analysis date	106 (September 1992)
Index at preparation of first estimate	128 (June 1995)

Adjustment for location:

Location factor for building analysed	1.00 (Reading)
Location factor for the proposed building	0.94 (Portsmouth)

Remember. The index has been updated to the time the estimate is being prepared – not to project completion.

The adjustment for time and location may be combined into one Adjustment Factor (*AF*) thus,

$$AF = \frac{128}{106} \times \frac{0.94}{1.00} = 1.1351 \quad \text{say} \quad \mathbf{1.14}$$

These adjustments have been calculated as a normal routine approach, accepting without question published indices and factors. It should not be forgotten that this approach should be combined with professional judgement to reflect personal knowledge of local conditions and current trends. For our example the other factors previously discussed under 'price level' do not apply.

Adjustment for Quantity and Quality

To a large extent, size differences between the office block analysis and the new project will be catered for by considering the cost per m^2 gfa. In addition we must make adjustments where major differences in the quantity or quality of an element can be found. At this stage, when considering quantity we must be alert to changes in ***densities*** (as discussed in Chapter 9) as this is most important when considering costs in terms of £/m^2 per gfa. Similarly, when examining quality, some thought should be given to the ***mix of individual items*** within an element (as discussed in Chapter 10, Section C). Given the limited information available, there is little to be gained by presenting the first estimate in an elemental format.

Inspection reveals a number of items in the analysis for which adjustments are required.

FS

1 Office partitions

The cost analysis includes demountable partitioning for office areas (see Appendix A). In the new project, apart from WC cubicles, this will be the tenant's responsibility. The reduction is as follows:

Demountable partitioning = £14 040

$$\frac{£14\,040}{1408\,m^2} = £9.97 \text{ per } m^2 \text{ gfa}$$

(1408 m^2 is the gross floor area of the building analysed.)

Update and adjust for location

$$£9.97 \times 1.14 = £11.37 \text{ per } m^2 \text{ gfa}$$

Floor finishes and suspended ceilings

The cost analysis shows that the carpet tiles were supplied under a separate contract and this supply cost is not included in the analysis (see Appendix A). Thus we must make the adjustment to the 'fix only' price for carpet tiles. We take the area of 1040 m^2 (the quantity shown in the analysis for carpet tiles). We decide, after looking through manufacturers' catalogues, upon a current rate of £13 per m^2 as an appropriate price for the supply of carpet tiles. We can calculate an appropriate addition per m^2 gfa for this adjustment.

$$\frac{1040\,m^2 \times £13}{1408\,m^2} = £9.60 \text{ per } m^2 \text{ gfa}$$

With regard to ceiling finishes, the cost analysis shows that plasterboard and skim coat on timber framing has been taken throughout the building analysed. For the proposed project, acoustic tiles in a suspended metal grid is required for the office areas. Plasterboard and skim will be retained for toilet areas.

We find a current price for suspended ceilings to our specification of £29.54 per m^2. The analysis shows the area of ceiling finish to be 1380 m^2. From this we calculate the price per m^2 gfa as follows:

$$\frac{1380\,m^2 \times £29.54}{1408\,m^2} = £28.95 \text{ per } m^2 \text{ gfa}$$

To calculate the extra cost we must deduct from this figure the cost (after updating and adjusting for location) included in the analysis for the plasterboard and skim finish (i.e. £14.07 × 1.14). Thus £28.95 – (£14.07 × 1.14) gives an extra cost of £12.91. This will be slightly on the high side, since suspended ceilings have now been allowed everywhere (including toilet areas). The additional cost is considered too small to worry about at this stage.

FS

3 Lifts

The cost analysis shows that there were no lifts in the analysed building (5J: Lift installations). The design team take the early decision that the building will be three storeys high. Although not absolutely essential, it has been decided that two lift cars will be installed in the new project. We must therefore allow for this element.

The design team give us their early impressions regarding specifications of the lifts. This includes type of lift, car size, finishes, travel speeds control system, position of lift motors, and the like. Armed with this information we seek the advice of a lift manufacturer and installer. Thus, it is established that a reasonable total cost will be £60 000. This is a current estimate as applicable to our new three-storey project and nothing will be gained by relating this figure back to our cost analysis. We still need a rate per m^2 gfa, so we use the 2250 m^2, i.e. the gross floor area of the new project.

$$\frac{£60\,000}{2250\,\text{m}^2} = £26.67 \text{ per m}^2 \text{ gfa}$$

By inspection of previous projects, we assess the builder's work in connection with the lifts at:

£3200 or £1.42 per m^2 gfa

4 Cost implications of three-storey building and increased storey height

In order to provide the best cost comparison our cost analyses should have been of three-storey buildings with appropriate storey heights for our new project. However, we are suggesting that we were unable to find suitable analyses. Therefore, we must now consider the cost implications that this might have for our feasibility estimate.

Very clearly, there will be implications for the densities of some elements and, by implication, the associated costs per m^2 gfa. For example, in our cost analysis, the ratio of upper floors to gross floor area is:

$$\frac{704}{1408} \quad \text{i.e. } 1:2$$

If an additional upper floor were to be added to the analysed building, gfa would increase by 704 m^2 giving a revised gfa of 2112 m^2 (1408 m^2 + 704 m^2). This would thus affect the density of this element and the ratio would change to:

$$\frac{1408}{2112} \quad \text{i.e. } 2:3 \quad (1:1.5)$$

Accordingly, the costs per m^2 gfa for this element would increase. In very approximate terms, with the addition of a third floor, the new element total cost would be doubled to £45 258. Then, using the new and increased gfa of 2112 m^2, the cost per m^2 gfa would equate to £21.43. (If you wish to review this topic, refer back to Chapter 9 where this concept was first introduced.)

FS

Following on from this, as each element is inspected in turn, we should find that some costs per m^2 gfa will increase while others will decrease.

We now intend to consider the cost implications of the three-storey building and increased storey height. As there are likely to be implications for most of the superstructure, we need to examine all the elements in this group. Remember, the information at this stage is limited and we therefore have to make adjustments to the analysis, rather than rely on measuring element quantities for the proposed project.

In order to simplify the process, all adjustments are made before any price level changes are carried out. The adjustment for price level is applied, at the end of this section (once only).

(a) Substructures

We ascertain from the structural engineer that the addition of the third storey and increased storey heights will have little effect on the foundation design as described in the cost analysis. Therefore we believe that a similar solution to that of the cost analysis may well suffice. However, the cost per m^2 gfa of substructures will now fall, as the ratio between gfa and the element unit quantities for substructures increases.

Cost per m^2 gfa in analysis = £32.13 m^2
Element unit rate in analysis = £64.27 m^2
Revised gfa if analysis were for three storeys = 2112 m^2

New cost per m^2 gfa:

$$\frac{704 \times £64.27}{2112\,m^2} = £21.42\,m^2$$

Reduction to the analysis rate is

£32.13 – £21.42 = £10.71 per m^2 gfa

(b) Frame

On first inspection it might be considered that increases in the cost of the frame will be adequately dealt with by the increased gross floor area of the new project. To a large extent this is correct. However, additional costs due to extra storey height have not been allowed for. We will make allowances for this in a rather crude way by increasing the cost per m^2 gfa in proportion to the differing storey heights.

FS

Storey height of analysed building = 2.75 m
Proposed storey height of project = 3.5 m (average)

The increase in storey height is 750 mm. This affects the costs of the columns. In fact, a quick approximate estimate, based upon the assumptions of 400 mm × 400 mm concrete columns on a three-metre grid, indicates minimal extra cost. However, the cost in the analysis of £32.22 m^2 gfa appears low when compared with other analyses. An appropriate range seems to be £40.00 to £65.00. We therefore decide to increase this element cost up to the low end of our range by adding 25%.

Cost per m^2 gfa in analysis = £32.22

25% × £32.22 = £8.06

Addition to analysis rate = £8.06 per m^2 gfa

(c) Upper floors

Similar arguments to those applied to substructures are also applicable to upper floors. However, this time the cost per m^2 gfa will increase.

Cost per m^2 gfa in analysis = £16.07 m^2
Element unit rate in analysis = £32.14 m^2
Revised rate per m^2 if analysis were for three storeys
i.e. two upper floors (704 × 2) = 1408 m^2

Converting to cost per m^2 gfa gives

$$\frac{1408 \times £32.14}{2112\,m^2} = £21.43\,m^2 \text{ gfa}$$

Addition to analysis rate is

£21.43 – £16.07 = £5.36 per m^2 gfa

(d) Roof

Adding a storey to the analysed building means that the same roof area is now covering an additional floor. Thus, reductions in the cost per m^2 gfa will be achieved.

Total cost in analysis = £64 491

Converting to cost per m^2 gfa:

$$\frac{£64\,491}{2112\,m^2} = £30.54 \text{ per } m^2 \text{ gfa}$$

Reduction to analysis rate is

£45.80 – £30.54 = £15.26 per m^2 gfa

FS

(e) Stairs
The stair requirements for the new project could be very similar to the provisions described in the analysis. The major difference being the additional travel from first to second floor. This will roughly double the quantities and cost contained in the analysis. Thus,

$$£15\,758 \times 2 = £31\,516$$

Also to be considered is an allowance to cover the increased storey height. We will adopt the same crude approach as we did for the frame and add 25% to our total cost of this element.

$$125\% \times £31\,516 = £39\,395$$

Converting to cost per m^2 gfa:

$$\frac{£39\,395}{2112\,m^2} = £18.65$$

Addition to analysis rate is

$$£18.65 - £11.19 = £7.46 \text{ per } m^2 \text{ gfa}$$

(f) External walls, and (g) Windows and external doors
We decide to take these two elements together as there are several factors common to both elements. The ratio of external wall to windows in the analysis is about 40 : 60. With the new project it is suggested that this ratio will be reversed with the external envelope composed of 85% external walls and 15% windows. With traditional brick construction, it is quite normal for the element unit rate of windows to be three to four times higher than those for external walls depending on the specification. Curtain walling, which is normally considered as part of the external wall element, is also relatively expensive. We would therefore expect to see a big reduction per m^2 gfa overall for these two elements when they are considered together as one. This will be so, even after allowing for the increased storey height.

Taking the analysis, total the two element quantities, increase by 25% (increase in storey height) and then split in the ratio of 15% windows : 85% walls. Thus

$$(531 + 304) \times 125\% = 1044\,m^2 \qquad 157 : 887$$

Then (using ratio):

External walls	$887\,m^2 \times £89.04 = £78\,978$
Windows and doors	$157\,m^2 \times £324.96 = £51\,019$

Thus, cost per m^2 gfa for the two elements:

in analysis $£19.23 + £122.55 = £141.78$ per m^2 gfa

FS

$$\text{when revised} \quad \frac{£78\,978 + £51\,019}{1408\,\text{m}^2} = £92.33 \text{ per m}^2 \text{ gfa}$$

Reduction to the analysis rate for these two combined elements is in the region of £141.78 – £92.33 = £49.45 per m^2 gfa

The reductions are possibly even larger than are suggested by these calculations. This is because the element unit rate for curtain walling has been used rather than a rate for windows, and this looks high. However, with the lack of information at this stage, we will accept this figure.

(h) Internal walls and partitions

We have already made adjustments for the demountable partitions. With these omitted from this element, the cost per m^2 gfa is as follows:

$$\frac{£26\,295 - £14\,040}{1408\,\text{m}^2} = £8.70 \text{ per m}^2 \text{ gfa}$$

At this stage, we can only assume the densities will be the same and thus solely allow for the increase in the storey height. We propose doing this as before by adding 25% to our rate. Therefore 25% addition is

$$25\% \times £8.70 = £2.18 \text{ per m}^2 \text{ gfa}$$

(i) Internal doors

We chose to make no change.

(j) Internal finishes

Floor finishes and ceiling finishes have already been adjusted so we need only to consider wall finishes. The change in the ratio of external wall area to window area will result in an increase to the area of wall finishes. Internal walls and partitions are unaffected by this change. To make allowance for this increase we propose to increase the cost in the analysis by 50%. The rate per m^2 gfa in the analysis is £13.44, therefore a 50% addition is £6.72, which is to be added to the analysis.

(k) Fittings

No adjustment required.

(l) Services

With increased storey heights, we will make the following assumptions:

- ❑ increases in vertical pipe runs,
- ❑ no increases in horizontal pipe runs,
- ❑ possibly a slight increase in the number of pipe fittings,
- ❑ certainly no increases in other fittings and fixtures (where much of the cost will be concentrated).

FS

To allow for this we will add an arbitrary 5% to the cost analysis rate per m^2 gfa:

£227.80 × 5% = £11.39 per m^2 gfa

We can now summarise the adjustments that we have made in connection with adding a third storey and increasing storey heights.

Element	£ per m^2
(a) Substructures	−10.71
(b) Frame	+ 8.06
(c) Upper floors	+ 5.36
(d) Roof	−15.26
(e) Stairs	+ 7.46
(f) External walls (g) Windows and external doors	−49.45
(h) Internal walls and partitions	+ 2.18
(i) Internal doors	n/c
(j) Internal finishes	+ 6.72
(k) Fittings	n/c
(l) Services	+ 11.39

The cost implications of a three-storey building with increased storey height = −£34.25 per m^2 gfa

In order to ensure that we deal with all adjustments in the same way, we must now adjust for price level to reflect current prices.

−£34.25 × 1.14 = −£39.05 per m^2 gfa

FS

We summarise the total cost of the office portion of our project (excluding external works, preliminaries, and contingencies):

			£ per m^2 gfa
Net cost from cost analysis (i.e building sub-total excluding preliminaries and contingencies)			612.22
Update and adjust for location (as we are dealing with costs current at date of estimate)		612.22 × 1.14 =	697.93
Less	Office partitions	11.37	686.56
Add	Floor finishes	9.60	
	Ceiling finishes	12.91	
	Lift installation	26.67	
	Builders work in connection with lifts	1.42	
Less	Cost implications of 3-storey construction, etc.	–39.05	
			£698.11

Multiply by gross floor area: £698.11 × 2250 m^2 = £1 570 747.50

Net cost of office portion = £1 570 750

It is worth comparing our cost of £698.11 per m^2 gfa (excluding preliminaries) with the histogram at the beginning of this section (Fig. 11.1). Early information suggests that the new project will be above the average quoted.

To complete our estimate, we now consider the following factors:

- External works. (This includes allowances for hard and soft landscaping, any work to road car parking and paved areas, fencing, work to site boundaries, and drains and sewer work. Water, electric, gas and sewerage charges include the cost of connection to the mains plus an assessment of distribution costs from the mains.)
- Preliminaries.
- Contingencies (i.e. a reserve against unforeseen items that occur during construction on site).
- Reserve against 'price and design risk (i.e. a reserve against possible price increases between the preparation of the estimate and the date of the tender).

After preparing approximate quantities, we decide that £180 000 is a suitable allowance for external works (the main consideration here has been car parking spaces).

FS

Allowing for preliminaries as a percentage of the project cost is not ideal. Further research and analysis of contractors' tenders and pricing methods is required. One approach would be to use cost significant items to build up an allowance for the cost plan. In most cases, this would involve no more than 10 or 12 cost significant items. However, for the sake of expediency we will continue to adopt the common method of inserting a percentage for this element. (Refer to any text in estimating to see how the preliminaries price is built up.)

After inspecting the figures included in the cost analysis for preliminaries and contingencies, we decide to increase both percentages slightly. Thus for preliminaries and contingencies we allow approximately 10% and 1.5% of the remainder of our estimate for these two items.

For the design risk at this early stage we decide to include 2.5%.

Now we need to consider price risk and insert a reserve against price rises which can occur between the date of producing this estimate at feasibility stage (June 1995) and the date for receipt of tenders (September 1996).

There are various forecasts of future cost trends available which would be consulted. Tender price forecasts should always be combined with professional judgement. The adjustment can then be applied as a percentage adjustment or as an adjustment factor, much in the same way as explained already. We have carried out the adjustment with a percentage addition.

FS

Summary – Feasibility stage estimate

	£	£
Office portion	1 570 750	
External works	180 000	
		1 750 750
Add		
Preliminaries (10%)	175 075	
Contingencies (1.5%)	26 261	
		1 952 086
Add		
Reserve against unforeseen design eventualities (2.5%)	48 802	
Reserve against price rises from June 1995 to September 1996 (approx. 5%)	97 604	
Total		**£2 098 492**

This estimate would probably be rounded off to say **£2 100 000** and then incorporated into the feasibility report presented to the client.

However, at feasibility stage, cost consultants do not always like reporting a single figure because design information is extremely limited. In this situation, one of the following two courses of action is usually taken:

(1) Where the client has set a cost limit, the report would confirm whether or not the project could be constructed within that limit.
(2) Where no cost limit has been set, a cost range might be reported together with some broad indication of quality.

(**Note:** *If you jumped to page 228 from Chapter 10, return to Chapter 10 after you have read this page; if not, continue to page 243.)*

We will assume that the client instructs the design team to proceed with outline proposals after he has considered the feasibility report.

The cost limit for the project is established at **£2 100 000** which relates to the proposed tender date of September 1996.

Outline proposals stage (September 1995)

Preparation of outline cost plan

The information available at this stage includes:

(1) Feasibility report, including the first estimate. (A cost limit has been established.)
(2) General brief and design criteria. (The design criteria are included here.)
(3) Town planning policy.
(4) Outline design drawings. In reality these would emerge during this stage but we have assumed that they are available for the production of the outline cost plan. These drawings have not been included here but we assume that they are similar to the scheme design drawings in Appendix B. In reality these would be less detailed and there would be much more uncertainty concerning measurements and dimensions. In this regard, you will have to accept the measurements presented in Table 11.1 and within the text.

The data used to estimate cost targets for the outline cost plan will be the same as at feasibility stage and are summarised below.

Table 11.1 Measures of quantity for the outline cost plan

Gross floor area	2250 m^2
Area of ground floor slab	750 m^2
Area of upper floors	1500 m^2
Area of roof (sloping area)	950 m^2
Area of external walls (including windows and doors)	1550 m^2

The gross floor area of our outline design (as measured from the outline design drawing) is approximately 2250 m^2. This gives us a total cost per m^2 gfa of

$$\frac{£2\,100\,000}{2250\,\text{m}^2} = £933.33 \text{ per m}^2 \text{ gfa}$$

Reserve for price and design risk

Before preparing our outline cost plan we should consider again our reserve against price and design risk.

At the feasibility stage, we allowed £146 406 representing a total of 7.5% of our estimated sum.

Our office block is seen as being a relatively straightforward project. Therefore, we decide that the allowance of 2.5% is a large enough margin for any unforeseen eventualities in the development of the design and thus leave this percentage unaltered.

We must now consider how prices are expected to move between the preparation of this outline cost plan (September 1995) and the receipt of tenders (September 1996). This is done by reference to appropriate forecasts. Following this approach, we decide to allow 4.5% to cover against the likely tender price rises over this period. Together this gives us a revised total design and price risk allowance of 7%.

Thus there will be little change on the figures produced for the first estimate. The total sum was:

£146 406 i.e. £65.07 per m^2 gfa

It will be helpful to keep this figure in mind when estimating cost targets for element groups. The actual price and design risk allowance will be calculated when the summary of the outline cost plan is produced. What has to be remembered is the cost limit which has now been set.

We prepare our outline cost plan by comparing the design criteria (as given on the next two pages) and the outline design drawing (similar to the scheme design drawings shown in Appendix B) with the cost analysis (Appendix A), making appropriate adjustments where necessary for major differences.

(*Note:* We clearly indicated at the feasibility stage a separate allowance for design risk; however, a sum of money for design risk is often not shown separately until the outline design stage simply because of the lack of definition in the earlier stage both in design detail and the estimating approach. Where this latter approach is adopted, then, the allowance for design risk should not be an addition to the estimate prepared at feasibility stage, but rather a reserve which is drawn from that total. The opposite is true for price risk as an allowance for this is usually calculated and shown separately from the outset.)

OP

Design criteria for office block

Site

(1) Site information and ground conditions

(a) Levels. Site plans would normally be available at this stage indicating relevant site features and indicating a grid of levels or the site contours.
(b) Soil. Clay and gravel.
(c) Bearing capacity. 200 kN/m^2.
(d) Water table. None encountered
(e) Depth of bearing strata. For three-storey block – 1.25 metres.
(f) General. Site cleared of old terrace houses without basements. Foundations grubbed up under a separate demolition contract.

General requirements

(2) The building

The building shall conform to the provisions of the current Building Regulations. In particular, the structure and protected shafts shall be of 1.5 hour fire resistance. Doors to protected shafts shall be of 1 hour fire resistance.

(3) Working conditions

Temperature, ventilation, lighting, sanitary conveniences, etc. shall comply with the latest requirements of current legislation.

Structure

(4) Frame

Storey height for all floors	3.5 m
Span	approximately 15 m
Fire resistance	1 hour

(5) Upper floors

Imposed load	5 kN/m^2
Fire resistance	1 hour

(6) Roof

Imposed load	0.75 kN/m^2

(7) Internal doors

Doors to protected shafts	1 hour fire resistance

OP

Environmental engineeering services

(8) Heating and ventilation

Air-conditioning using four-pipe fan coil comfort cooling system.

No. of air changes per hour	3
Temperature	22°C (dry bulb)
	50% RH

(9) Electrics

Lighting standard 500 lux

(10) Other services

All normal services to typical commercial standards including:

- hot and cold water supply
- fire services
- small power
- provision for communications
- lightning protection
- lifts
- drainage, etc.

Substructure (i.e. work below lowest floor finish)

The engineer advises that there is no significant soil problem, so a similar solution to that of the cost analysis may well suffice. It is assumed that the loads imposed by an additional storey plus the increased storey height will not alter the foundation design described in the cost analysis. There is a small addition for foundations under the lifts, but this is not a major cost. The existing foundations have already been grubbed up and no major problems are envisaged here. On reflection and after comparison with other projects, we decide the rate in the cost analysis is acceptable.

We therefore update the element unit rate from the cost analysis (i.e. £64.27 m^2) making the adjustment for location and applying the revised rate to our outline design; this is done by reference to suitable indices.

Adjustment Factor (AF)

Index at analysis tender date = 106
Index at preparation of outline cost plan = 130

(*Note:* The index changed from 128 to 130 between June 1995 and September 1995.)

Location factor for Reading (analysed building) = 1.00
Location factor for Portsmouth (new project) = 0.94

$$AF = \frac{130}{106} \times \frac{0.94}{1.00} = 1.15283 \quad \text{say} \quad 1.15$$

Update element unit rate and adjust it for location:

$$£64.27 \times 1.15 = £73.91 \text{ per m}^2$$

Area of ground floor slab measured from the outline design drawing = 750 m^2.

Allowance for substructure = 750 m^2 ground floor slab @ £73.91 per m^2

Cost per m^2 gfa:

$$\frac{750 \times £73.91}{2250\,\text{m}^2} = \mathbf{£24.64\ per\ m^2}$$

Superstructure

We shall now examine the elements individually. The major impact will be due to the additional storey and increased storey height.

(a) Frame

The cost analysis indicates a traditional steel frame to accommodate steel decking and lightweight in situ concrete upper floors. The costs of fire protective casing to central and exposed columns have been included in this element. Also included in this element are the costs of isolated beams located in the plant room. We will probably continue to use the same type of frame design as that employed in the cost analysis.

The cost included in the analysis is £32.22 per m^2. On initial inspection this rate appears low. In any event, it should be increased in line with the differences indicated at the feasibility stage. The main differences, as already identified, are the additional storey, an increase in the gross floor area, plus the increase in storey height. The total cost for the frame needs to be increased accordingly. Examining frame costs on other project analyses shows this element cost to be in the range of £40.00–£70.00 per m^2 gross floor area. (*Note:* These are current costs.) Some cost variability can be attributed to the type of protection used and whether this cost has been included here or recorded under 5K. Protective installations.

OP

After a further detailed inspection of our data source we identify the most appropriate analyses. From these we settle upon a cost of £60.00.

Gross floor area	$= 2250\,m^2$
Cost of element frame = 2250 × £60.00	= £135 000
Cost per m^2 gfa	= £60.00

(b) Upper floors

The cost analysis is for a two-storey building and our outline design is for three storeys. We must obviously allow for the additional upper floor. In addition the cost analysis describes a composite upper floor construction combining steel decking with a lightweight concrete topping. It is quite possible that our design team will select a different solution, but at this stage we do not believe the costs will be very different. (We will check this at detail design stage when more details are available.)

Update element unit rate and adjust it for location:

$$£32.14 \times 1.15\ (AF) = £36.96 \text{ per } m^2$$

(For calculation of Adjustment Factor (*AF*), see page 247.)

Calculate element cost from the outline design drawing:

Allowance for upper floors $= 1500\,m^2$ a £36.96 per m^2 = £55 440

Cost per m^2 gfa:

$$\frac{£55\,440}{2250\,m^2} = £24.64 \text{ per } m^2 \text{ gfa}$$

(c) Roof

We have assumed the design and specification will be similar. Adding an additional storey will reduce the cost of the roof when expressed in terms of £ per m^2 gfa.

Update element unit rate and adjust it for location:

$$£71.66 \times 1.15\ (AF) = £82.41 \text{ per } m^2$$

Calculate element cost from the outline design drawing (measured as sloping area of roof):

Area of roof $= 950\,m^2$

Allowance for roof $= 950\,m^2$ @ £82.41 per m^2 = £78 290

Cost per m^2 gfa:

$$\frac{£78\,290}{2250\,m^2} = £34.80 \text{ per } m^2 \text{ gfa}$$

(d) Stairs

The analysis shows two internal staircases constructed in concrete and two external escape staircases in galvanised steel, spiral pattern. The same requirements with regard to number and construction will suit the new project as can be seen from the outline design. However, what is different is the extra overall height made necessary by the additional floor and the increase in storey height.

From the outline design drawings we estimate that the rise of the staircase from ground to second floor is 7.00 m. This compares with 2.75 m (ground to first floor) in the analysed building. We will now use this information as a crude factor to adjust the total cost of the element. At this stage we see no need to isolate the costs associated with the two types of staircase design.

From the analysis, total cost of element = £15 758

$$\text{New total for element} = £15\,758 \times \frac{7}{2.75} = £40\,111$$

Update this element total cost and adjust it for location:

$£40\,111 \times 1.15\ (AF) = £46\,128$

Reduce to cost per m² gfa:

$$\frac{£46\,128}{2250\ \text{m}^2} = £20.50 \text{ per m}^2 \text{ gfa}$$

(e) External walls

The external envelope of the analysed project has a greater area of windows than our new project. Although the quality of the brickwork is very similar, there are fewer straight runs and more intricate work around column casings. Due to these characteristics we assess the element unit rate in the analysis to be slightly on the high side. Therefore we have made additional reference to other analyses; and selected a current rate appropriate to the location of £80.00.

We measure the external wall area on our project from the outline design drawings by assuming a distribution of 85% solid wall and 15% windows and doors; this gives us an external wall area of 1318 m².

Element unit cost:

$1318\ \text{m}^2\ @\ £80.00 = £105\,440$

Cost per m² gfa:

$$\frac{£105\,440}{2250\ \text{m}^2} = £46.86 \text{ per m}^2 \text{ gfa}$$

OP

(f) Windows and external doors

Glazing style differs significantly between the analysis and the new project. Other analyses are used to select a rate appropriate to the outline design. The design team indicate that a target cost for above-average components should be allowed. Accordingly, we selected a current rate appropriate to the location of £330.00.

Following on from our calculations of external walls, the window and door area is assessed at 232 m^2 by assuming 15% windows and doors.

Element unit cost:

$$232\,m^2 \text{ @ £}330.00 = \text{£}76\,560$$

Cost per m^2 gfa:

$$\frac{76\,560}{2250\,m^2} = \text{£}34.03 \text{ per } m^2 \text{ gfa}$$

(g) Internal walls and partitions

It was difficult to measure the element unit quantities. After some consideration, we decided to adopt the quality and proportional quantity in the analysed building after allowing for the demountable partitioning to the office areas which is not required. (Again at this stage we have made allowance for the increases in internal wall heights due to the increase in storey height.) We calculate what the total cost per m^2 gfa would have been for internal walls and partitions in the analysed building if it had not provided demountable office partitions, adjust for increased height, then update to obtain a realistic cost per m^2 for our project.

Total cost of internal walls and partitions in cost analysis = £26 295

Less

Demountable partitions = £14 040

£12 255

£12 255 × 125% = £15 319

$$\frac{\text{£}15\,319}{1408\,m^2 \text{ (gfa in analysis)}} = \text{£}10.88 \text{ per } m^2$$

Update and adjust for location:

$$\text{£}10.88 \times 1.15\ (AF) = \text{£}12.51 \text{ per } m^2 \text{ gfa}$$

(h) Internal doors

We decide that we can adopt the quality and proportional quantity in the analysed building.

Cost per m^2 gfa = £5.87

Update and adjust for location:

$$\text{£}5.87 \times 1.15\ (AF) = \text{£}6.75 \text{ per } m^2 \text{ gfa}$$

We must now add these eight allowances to obtain our allowance for the superstructure.

Element	£ per m^2
(a) Frame	60.00
(b) Upper floors	24.64
(c) Roof	34.80
(d) Stairs	20.50
(e) External walls	46.86
(f) Windows and external doors	34.03
(g) Internal walls and partitions	12.51
(h) Internal doors	6.75
Total allowance for superstructure	**240.09 per m^2**

Internal finishes

The choices and standards of internal finishes for the analysed building are set out in the specification and design notes. Apart from one or two exceptions, these are quite suitable for the new project. So we will accept the analysis figure after adjusting for the supply of carpet tiles and allowing for an increase in the quantity of wall finishes and quality of the ceiling finishes. After checking the current rates again there is no reason to change the allowances made during the feasibility stage. So we will add £9.60 per m^2 gfa for the carpet tiles, and £12.91 per m^2 gfa for upgrading the ceiling finishes. We use the same approach as at feasibility stage to allow for the change in the ratio of windows to external walls. Thus, we add £6.72 for the increased quantities of wall finishes.

From the analysis,

		£ per m^2
Current cost of internal finishes = £73.43 × 1.15 (*AF*)	=	84.44
Increased quantity of wall finishes = £6.72 × 1.15 (AF)	=	7.73
Add allowance (all at current rates) for:		
Supply of carpet tiles	=	9.60
Suspended ceiling finishes		12.91
		£114.68 per m^2

Fittings

We require very few fittings in our project since these will be mainly the tenants' responsibility. After inspection, we decide to keep the same rate of £7.24 per m^2 gfa as in the analysis.

Update and adjust for location:

£7.24 × 1.15 = **£8.33 per m^2**

OP

Services

At feasibility stage we based our estimate for services on the figure contained in the cost analysis. We added 5% to the rate as an allowance for the extra vertical pipe runs associated with the increase in storey height. In addition we also included the costs for the lift installations together with the extra builders' work in connection with the lifts.

No other design decisions have been made during outline proposals which make the feasibility allowances for these items invalid (see pages 234 and 238). We will use the same allowances here.

From the analysis,

		£ per m²
Cost of services	=	227.80
Add:		
Extra vertical pipe runs		11.39
		239.19
AF × 1.15	=	275.07
Allowance for lift installation	=	26.67
Builder's work in connection with lifts	=	1.42
		£303.16 per m²

External works

From our original assessment used at the feasibility stage, we decide that the allowance for external works can remain unaltered. (We are obviously ignoring any slight increase, which may have occurred since we prepared this first estimate.)

$$\frac{\pounds 180\,000}{2250\,\text{m}^2} = \mathbf{\pounds 80.00\ per\ m^2}$$

Preliminaries and contingencies

After due consideration, we decide that we can allow for preliminaries and contingencies at the same rates as we chose at feasibility (i.e. 10% and 1.5% of the remainder of the outline cost plan).

Summary – Outline cost plan

	£ per m^2	
Substructure	24.64	
Superstructure	240.09	
Internal finishes	114.68	
Fittings	8.33	
Services	303.16	
External works	80.00	770.90
Add		
Preliminaries (10%)		77.09
Contingencies (1.5%)		11.56
		859.55
Add		
Reserve against unforeseen design eventualities (2.5%)		21.49
Reserve against price rises from September 1995 to September 1996 (4.5%)		38.68
Total		**£919.72 per m^2**

£919.72 × 2250 m^2 = £2 069 370.00

(as at anticipated tender date of September 1996)

The design team note how close it is to the feasibility stage estimate of £2 100 000, and accordingly report no change to the client, confirming that the cost limit still holds good.

SD

Scheme design stage (November 1995)

It is assumed that the client accepted the first estimate between feasibility and outline proposals: we therefore regard our first estimate (£2 100 000) as being the cost limit for the project. During the outline proposals stage, we prepared our outline cost plan showing the major cost allocations and confirmed that the cost limit was reasonable.

We will now prepare our detailed cost plan. This will show how we propose to distribute expenditure among all the elements of the project. It should be remembered that this cost plan is a reflection of the client's needs and the design team's choice of standards and specification. Another client or a different design team might produce a very different cost plan.

Another design team may choose to differ from us in the choice of method for preparing cost targets for each element. Consultants may prefer to use approximate quantities or inspection where we use simple proportion, and vice versa. This is not a matter of great importance, as long as we achieve our objective of producing a cost plan which allocates the available money in such a way that the design team's wishes are satisfied. That is, sufficient money is allocated to each element to enable the desired standards to be achieved.

The information available at this stage includes:

(1) Report produced at the end of the outline proposals stage, together with the outline cost plan.
(2) Design criteria and complete brief (not included here but assumed).
(3) Scheme design drawings included in Appendix B. In reality these drawings would be produced during this period and more information than included here would be available.
(4) The measures of quantity (element unit quantity) calculated from these drawings (see next page).
(5) A record of all the decisions made to date between client and design team.

Although difficult to convey in book form such as this, you should be aware that design and project information increases as the design develops. Thus the measured quantities can be reasonably accurate, although not yet to the stage of detailed measurement. The figures are therefore a reflection of this increase in information.

We shall adopt the same procedure as at the outline proposals stage and assume that the design information produced during scheme design has all been assembled before we start to prepare the cost plan.

Table 11.2 Measurement of quantity for the detailed cost plan

Gross floor area	2246 m^2
Area of ground floor slab	749 m^2
Area of upper floors	1498 m^2
Area of roof (sloping area)	957 m^2
No. of flights of stairs (internal stairs)	12
No. of external fire escape stairs	2
Area of external walls	1328 m^2
Area of windows	213 m^2
Area of external doors	12 m^2
Number of single internal doors	30
Number of double internal doors	6
Area of floor finishes	1994 m^2
Area of ceiling finishes	1983 m^2
No. of sanitary fittings	62

To save space, the table includes only the 'quantities' required for the elements whose cost targets are prepared by simple proportion

Reserve for price and design risk

Our first action in allocating the cost limit is to reconsider the reserve against price and design risk.

Perhaps the most important of these is the probability that prices will rise between the preparation of the cost plan and receipt of tender. We review our prediction made at outline proposals stage (September 1995). This was based on the published indices, particularly the Tender Price Index (TPI) and forecast, modified by our professional judgement. Less than three months have passed since we last looked at this and no change has been noted. We will therefore maintain our allowance of 4.5% for tender price rises, that is, from November 1995 to September 1996.

We now turn our attention to the design risk element. The design has advanced since the outline cost plan was prepared and several of the major questions then outstanding have now been settled. However, important design questions are still outstanding which will be addressed during the detailed design stage. As a result, we decide that the allowance of 2.5% included at outline proposal stage should remain. Thus,

total reserve for price and design risk = 7%

The allowance included at the outline proposals stage was £60.17 per m^2 of gfa. This figure can be used as a guide when allocating costs between elements. The actual amounts we will include for price and design risk can be calculated later on our summary sheet.

We will now consider preliminaries and contingencies. When we have made allowances for these two important elements, we can consider the remainder of the elements in the normal order as indicated by our analysis (Appendix A).

Element 7 Preliminaries (including insurances)

The likely duration of the project and specific requirements for site staff, site accommodation, mechanical plant (particularly cranes), insurances, and the like should be considered at this point. Preliminaries in the analysed building when expressed as a percentage of remainder of contract sum (less contingencies) is 9.5%. We decide that the building analysed and our project are reasonably similar and consider that it is not necessary to make any alterations to the percentage allowed at the feasibility stage.

Following from this we assume that, since the percentage in the analysis appears reasonable, it is likely that none of the costs usually included in preliminaries have been distributed among the other elements. Similarly, after inspection of the analysis, we conclude that the contractor's profit has been spread evenly throughout the tender and is not loaded unevenly upon one or more elements. Therefore, if we adopt a similar percentage we can operate directly on all the other element costs in the analysis, safe in the assumption that they include profits, but exclude all the costs which properly belong to preliminaries.

Because of this, we will adopt the same allowance used in the outline cost plan (i.e. 10% or £77.09 per m^2).

Total allowance for preliminaries = £77.09 × 2246 = £173 144

	Total	Cost per m^2 gfa
Cost target for preliminaries	**£173 144**	£77.09

(*Note:* As we have used a percentage to calculate this element, we can anticipate that this cost target may change slightly when we calculate our final summary sheet.)

Contingencies

It is also convenient to consider contingencies at this point. In view of the straightforward nature of this project and this site, we recommend (and the client accepts) the allowance which we made during outline proposals (i.e. £11.56 per m^2).

Total allowance for contingencies = £11.56 × 2246 = £25 964
Round off to say £26 000

Cost per m^2 will therefore be

$$\frac{£26\,000}{2246} = £11.58$$

	Total	Cost per m^2 gfa
Cost target for contingencies	**£26 000**	**£11.58**

Element 1 Substructures

A comparison of the quantities and the design criteria for the analysed building and our project reveals the following significant differences:

- our project is three storeys high whereas the other was only two storeys;
- both buildings have structural steel frames, although the new project also has a significant amount of traditional brick external walling, built off a strip foundation;
- the number of column bases is greater in our proposed project than that given in the analysis
- the ground conditions for the analysed building were more difficult than for our project.

The structural engineer is of the opinion that, given the amount of external walling and when the different ground conditions are taken into account, economies could be incorporated into the design of the frame. We are satisfied that these economies will offset, to some degree, the increased number of column bases. It seems likely that the design costs for the two projects will be very similar. We decide, therefore, to deal with this element by simple proportions rather than resorting to approximate quantities.

Adjustment for Price

Adjustment Factor (*AF*)

Index at analysis tender date	106
Index at preparation of cost plan	130
Location factor for Reading (analysed building)	1.00
Location factor for Portsmouth (new project)	0.94

$$AF = \frac{130}{106} \times \frac{0.94}{1.00} = 1.15283 \quad \text{say} \quad 1.15 \quad \text{(as before)}$$

Update element unit rate and adjust it for location:

$$£64.27 \times 1.15 = \mathbf{£73.91\ per\ m^2}$$

Adjustment for Quantity

Current element unit rate	$= £73.91$ per m^2
Quantity (from scheme design drawings)	$= 749\ m^2$

$$749 \times £73.91 = £55\,359$$

Reduce to the common yardstick of cost per m^2 gfa:

$$\frac{£55\,359}{2246\ m^2} = \mathbf{£24.65\ per\ m^2}$$

(*Note:* The index for November 1995 (date of this detailed cost plan) has not changed from the index of September 1995 (date of outline proposal stage), indicating that tender prices have remained relatively stable over this period.)

Adjustment for Quality

We will make no adjustment for quality because of our assumption that the additional column bases will be offset by the saving that should result from improved ground conditions.

(*Note:* This assumption must be confirmed early in the cost checking process. At this point in time we are confident that any possible increases in price can be covered by the design risk reserves. If the differences between the analysis and the project had been more complex, we would have prepared this cost target by approximate quantities.)

	Total	Cost per m^2 gfa
Cost target for substructures	**£55 359***	**£24.65***

* Small discrepancies will appear due to the approximations made in conversions to cost per m^2 of gfa (i.e. £24.65 × 2246 m^2 = £55 364). These are ignored here and on the summary when preparing the cost plan.

Element 2A Frame

The analysed project has a traditional steel frame, supporting lightweight suspended floors on steel decking. The frame for our design is also traditional, but we have an extra floor and the storey heights are increased to accommodate suspended ceilings and raised floors. Also, as identified under substructure, there is a large increase in the number of columns. In addition it is thought likely that the specification for the upper floors will be changed to precast pre-stressed composite floors with in situ concrete topping (allowed for in the upper floors element).

These changes rule out adjustment by simple proportion and the nature of our frame rules out inspection, so we use approximate quantities. We show below the summary only, as the detailed dimensions are too lengthy to include here.

Summary

Fabricated structural steel framing (Grade 50C)

	Amount	Cost per unit	Total
Columns including connections	84 t	£900	£75 600
Beams including connections	18 t	£1000	£18 000
Roof trusses	13 t	£1250	£16 250
Bases, including wedges and anchorage	38	£250	£9 500
Erection	122 t	£150	£18 300
Treatment	122 t	£85	£10 370
			£148 020
Sundries 5%			£7 401
Total			£155 421

Reduce to cost per m² gfa:

$$\frac{£155\,421}{2246\,\text{m}^2} = £69.20\ \text{per m}^2$$

(*Note:* Another element requiring confirmation early in the cost checking process.)

	Total	**Cost per m² gfa**
Cost target for frame	**£155 421**	**£69.20**

Element 2B Upper floors

As explained when discussing the frame, the design of the analysed building and our new building are not alike.

The dissimilarity rules out simple proportion and the nature of the task rules out inspection, so again we use approximate quantities. This gives us a sum of £67 410. (The details of the calculations are too lengthy to include here.)

Cost per m^2 gfa:

$$\frac{£67\,410}{2246\,\text{m}^2} = £30.01\ \text{per m}^2$$

	Total	Cost per m^2 gfa
Cost target for upper floors	**£67 410**	**£30.01**

(*Note:* Our estimated cost targets for the last two elements, frame and upper floors, have increased significantly from the allowances made at outline proposals stage. If these increases are not balanced out by the estimated costs of the following elements then our cost limit will be exceeded. This will then entail a major review by the design team before reporting any findings to the client.)

Element 2C Roof

There appears to be many similarities between our new project and the analysed building, so we decide to use the simple proportion method to prepare our cost target.

Adjustment for Price

Element unit rate from analysis = £71.66 per m^2
Current element unit rate including adjusting for location, i.e.

£71.66 per $m^2 \times 1.15$ = £82.41 per m^2

Adjustment for Quantity

Quantity (from drawings)	= 957 m^2
Total element cost is 957 m^2 @ £82.41 per m^2	= £78 866

Cost per m^2 gfa:

$$\frac{£78\,866}{2246\,m^2} = £35.11 \text{ per } m^2$$

Adjustment for Quality

The structural steelwork in the roof has been included with element 2A (frame). This was also the approach adopted for the cost analysis. With regard to quality we need therefore to consider the coverings. The design team confirm that artificial slates, as used in the analysed building, are an adequate solution.

Therefore, we do not need to make any adjustments for quality.

	Total	**Cost per m^2 gfa**
Cost target for roof	**£78 866**	**£35.11**

Element 2D Stairs

No major change in design is envisaged, so we should obviously investigate the possibility of using simple proportion. No element unit rate is given in the analysis (see Appendix A). However, very often it is possible to use the reasonable assumption that the cost of a staircase is proportional to the number of flights (provided they are of the same size and quality).

The concrete staircase in the analysed building has two flights per staircase per floor plus a half landing (a total of four flights for the building). In the new project there are three flights per floor with two quarter space landings (a total of twelve flights for the building). Bearing in mind the additional storey height, this appears reasonable and suggests the flights on average might be of a reasonably similar size. We cannot apply the same approach for the escape stairs but the overall proportions do not seem too dissimilar. We therefore decide to make our adjustment based on number of flights and applied to the total element costs as if staircases were of a similar pattern and all constructed in concrete.

Adjustment for Price

Update element total cost and adjust for location:

$$£15\,758 \times 1.15 = £18\,122$$

Adjustment for Quantity

In the ratio of 12 : 4 (as explained above)

$$£18\,122 \times \frac{12}{4} = £54\,366$$

Cost per m^2 gfa:

$$\frac{£54\,366}{2246\,\text{m}^2} = £24.21\ \text{per m}^2$$

Adjust for Quality

The design team have indicated that they would prefer a better quality handrail and balustrade. We could attempt to evaluate such improvements by deciding on a particular specification for the balustrade (e.g. stainless steel and plate glass) and price this by approximate quantities. Alternatively, we could decide to increase the allowance for stairs by a certain percentage and to explore during the detail design stage exactly what detailed specification of balustrade we can afford. We will use this latter method. Bearing in mind the cost limit and the requirement of other elements, we decide to increase the allowance by 15% to improve the standard.

Former allowance = £54 366

Add increase of 15%:

115% × £54 366 = £62 521

Cost per m^2 gfa:

$$\frac{£62\,521}{2246\,\text{m}^2} = £27.84 \text{ per m}^2$$

	Total	Cost per m^2 gfa
Cost target for stairs	**£62 521**	**£27.84**

Element 2E External walls

The elements 2E (external walls) and 2F (windows and external doors) are examined together. This reveals that the quantities shown in the analysis are very different to our new project. The differences are mainly due to increases in the storey heights and changes in the ratio of windows to walling. Changes in specification are also noticeable. Double glazed units are likely to be used for windows in the new project. Although brickwork in the new project is generally less intricate, brick arches have now been used as a design feature over all ground floor window and door openings.

Therefore, it is apparent that adjustment by proportion is not an appropriate technique: we shall resort to approximate quantities based upon an assumed specification for both these elements. We shall assume that a reasonably high quality external facing brick will be required and therefore allow a prime cost of £525 per 1000 in our rate build up. In all other respects the external wall details will be of a normal standard for this type of project (blockwork inner skins and insulated cavities). The area of external walls as measured from drawings is 1328 m^2.

In summary, the approximate quantities calculations are as follows:

1328 m^2 of external walls @ £77.50 = £102 920

Cost per m^2 gfa:

$$\frac{£102\,920}{2246\ m^2} = £45.82\ \text{per}\ m^2$$

	Total	Cost per m^2 gfa
Cost target for external walls	**£102 920**	**£45.82**

Element 2F Windows and external doors

As previously decided, we will use approximate quantities to adjust this element. (The details of the calculations are too lengthy to include here.)

Summary of the approximate quantities calculations are:

$213\,m^2$ of windows and external doors @ £325 = £69 225

Reduce to cost per m^2 gfa:

$$\frac{£69\,225}{2246\,m^2} = £30.82 \text{ per } m^2$$

	Total	Cost per m^2 gfa
Cost target for windows and external doors	**£69 225**	**£30.82**

Element 2G Internal walls and partitions

The building analysed included costs for demountable partitions. In our project most of the partitioning is left as the tenant's responsibility. In addition, our standards are somewhat different and the increased storey heights must be taken into account: we therefore decide to use the approximate quantities technique.

73 m^2 of 100 mm blockwork partition @ £21.00	=	£1533
255 m^2 of 200 mm blockwork partition @ £32.00	=	£8160
77 m^2 of 112.5 mm common brick internal walls @ £23.00	=	£1771
18 WC cubicles @ £330	=	£5940
		£17 404

Cost per m^2 gfa:

$$\frac{£17\,404}{2246\,m^2} = £7.75 \text{ per } m^2$$

	Total	Cost per m^2 gfa
Cost target for internal walls and partitions	**£17 404**	**£7.75**

Element 2H Internal doors

The offices in the building analysed were fitted with similar specification flush doors. However, the proportion of single to double flush doors will be different to our project. Because of this, we cannot adopt the element unit rate from the analysis. However, since we envisage that similar types of door will be used, we can adopt the all-in unit rates (see Appendix A) for the flush doors.

Adjustment for Price

All-in unit rates from analysis are:
flush doors, including perimeter treatments, e.g. lintels, linings/frames, ironmongery, decorations, etc.

single doors = £240 each
double doors = £395 each

Updated all-in unit rates, adjusted *AF*

single doors £240 × 1.15 = £276 each
double doors £395 × 1.15 = £454 each

Adjustment for Quantity

The number of single and double doors, measured from the scheme design drawings, are

single doors = 30 nr
double doors = 6 nr

cost of single doors = £276 × 30 = £8 280
cost of double doors = £454 × 6 = £2 724
£11 004

Cost per m^2 gfa:

$$\frac{£11\,004}{2246\,m^2} = £4.90 \text{ per } m^2$$

Adjustment for Quality

Both the fire resistance standards of the internal doors and the quality of the doors and linings in the analysis are satisfactory for our project, so no adjustment is made for quality.

	Total	**Cost per m^2 gfa**
Cost target for internal doors	**£11 004**	**£4.90**

Element 3A Wall finishes

We decide that it is best to use approximate quantities based upon an assumed specification of plastered walls, with glazed tiles in the lavatory areas.

1689 m^2 of plaster and emulsion paint @ £15.00	=	£25 335
220 m^2 of plaster and fabric wall covering @ £27.00	=	£5 940
242 m^2 of glazed tiling @ £55.00	=	£13 310
		£44 585

Cost per m^2 gfa:

$$\frac{£44\,585}{2246\,m^2} = £19.85 \text{ per } m^2$$

	Total	Cost per m^2 gfa
Cost target for wall finishes	**£44 585**	**£19.85**

(*Note:* This does not commit us to final use of this specification, but simply to using one of a similar general quality. At this stage, we are making a reasonable allowance within which wall finishes of the desired quality can be designed during the detail design stage.)

Element 3B Floor finishes

Office floors in the analysed building are raised access with carpet tiles. The quality of the existing access floor is considered appropriate for our design. There must be an allowance for the carpet tiles, which in the analysed building were supplied free of charge from the client. The design team have indicated that a carpet tile of medium quality would be appropriate (current costs are about £13.00 per m^2); we find the rate used for the feasibility estimate still holds good.

We can use simple proportion for this element. However, remember that the carpet tiles do not cover the whole floor area. Therefore, we cannot just add the £13.00 per m^2 to the updated element unit rate. Instead we shall calculate what the total element cost would be if a building of exactly the same design as the analysed building were to be built now, with carpet tiles supplied and laid as floor finishes. From this we can find the element unit rate for the hypothetical building to use for our project. This approach assumes that our area of access floors with carpeted floor finish will be the same proportion of the total floor finishes areas as in the analysed building.

Adjustment for Price

Total element cost in analysis	£64 650
Updated element cost, adjusted for location £64 650 × 1.15	£74 348

Adjustment for Quality

Area covered by carpet in analysis is 1040 m^2, add for supply @ £13.00 per m^2	£13 520
Current cost of the element as specified in the analysis, plus supply of carpet	£87 868

Adjustment for Quantity

Total area of floor finishes in analysis = 1301 m^2
Total area of floor finishes in our project = 1994 m^2

$$\text{Our total element cost} = \frac{£87\,868 \times 1994}{1301} = £134\,672$$

Cost per m^2 gfa:

$$\frac{£134\,672}{2246\,m^2} = £59.96 \text{ per } m^2$$

	Total	Cost per m^2 gfa
Cost target for floor finishes	**£134 672**	**£59.96**

Element 3C Ceiling finishes

Plasterboard and skim coat on timber framing has been used throughout the analysed building. For the proposed project, acoustic tiles in a suspended metal grid are required for the office areas. Plasterboard and skim will be used for toilet areas.

Preliminary studies indicate that we can expect to obtain a high quality suspended ceiling for £34.30 per m^2. An appropriate rate for plasterboard and skim on softwood framing would be £16.78 per m^2. These rates are slightly higher than those considered at feasibility stage.

Area of ceiling finishes (from drawings) is
1888 m^2 suspended ceiling @ £34.30 = £64 758
95 m^2 plasterboard and skim @ £16.78 = £1 594
£66 352

Cost per m^2 gfa:

$$\frac{£66\,352}{2246\,m^2} = £29.54 \text{ per } m^2$$

	Total	Cost per m^2 gfa
Cost target for ceiling finishes	**£66 352**	**£29.54**

Element 4A Fittings and furnishings

This is not a major element of the project. Therefore the technique of inspection is used, for which we obtain a selection of suitable cost analyses to give us a range of element costs. This range is studied in conjunction with available information before making a final choice. We know that all the office fittings will be supplied by the tenant, so no provision need be made for these. However, it is prudent to make a small allowance for minor items, such as shelving, pinboards, mirrors, handicap toilet grab rails, signs, and perhaps louvre blinds, as was the case with the analysed building. After studying our range of costs we decide that a sum of £20 000 will be allowed for fittings.

Cost per m^2 gfa:

$$\frac{£20\,000}{2246\,\text{m}^2} = £8.90 \text{ per m}^2$$

	Total	Cost per m^2 gfa
Cost target for fittings and furnishings	**£20 000**	**£8.90**

Element 5A Sanitary appliances

This element may be handled by any of the three techniques, but we shall use approximate quantities.

18 no. low level WCs @ £230	=	£4140
12 no. suspended urinals @ £200	=	£2400
26 no. lavatory basins @ £180	=	£4680
6 no. cleaners' sinks @ £220	=	£1320
Short connections and overflows	=	£1820
		£14 360

Reduce to cost per m^2:

$$\frac{£14\,360}{2246\,m^2} = £6.39 \text{ per } m^2$$

	Total	Cost per m^2 gfa
Cost target for sanitary appliances	**£14 360**	**£6.39**

Note: It is interesting to compare the target cost of £14 360 with the sum obtained by simple proportion, making the Quantity and Price adjustments, (and ignoring any Quality adjustment), using element unit quantities based on the number of fittings in the analysis and in the new project:

Update element cost and adjust it for location:

$$£8110 \times 1.15\ (AF) = £9327$$

Adjustment for Quantity

$$\frac{£9327 \times 62}{38} = £15\,218$$

The outcome is obviously dependent on there being some relationship between the number of fittings required per m^2 of floor space, the type and the quality. Fundamentally it comes down to considering possible relationships between the planning parameters and occupants' requirements. For instance, office floor space will vary from between say $6\,m^2$ per person and $20\,m^2$ per person, depending on whether the design is for a high-density open-plan layout or a low-density cellular layout.

Element 5C Disposal installations

An unimportant element in this project which is best decided by inspection. (Approximate quantities could also be used if the money involved justifies the time required for measurement.)

The environmental services in our project are located in close proximity around staircase areas, thus reasonably short horizontal pipe runs can be expected. We need to remember that increased storey heights reflect the incorporation of suspended ceilings and raised floors in our design. These points must be considered when inspecting similar cost analyses and other data sources.

The costs of disposal installations in existing offices are as follows:

Building	Tender date	Preliminaries (%)	Cost of disposal installations (£/m^2 gfa)
2-storey offices *Winchester*	March 1991	10.27	3.50
4-storey offices *Manchester*	February 1993	7.66	4.86
2-storey offices *Slough*	January 1993	8.79	4.01
2-storey offices *Reading*	September 1992	9.5	3.89

Making approximate adjustments for the differences in the preliminaries percentages and the tender dates of these projects, we decide to adopt an allowance of £4.75 per m^2 for the element which includes for builder's work to pipe ducts.

Total element allowance = £4.75 × 2246 m^2 = £10 669

	Total	**Cost per m^2 gfa**
Cost target for disposal installations	**£10 669**	**£4.75**

Element 5D Water installations

Mains supply, cold-water services and hot-water services must all be considered with this element. The mains supply from point of entry up to storage tanks in the roof space will be similar to that in the cost analysis (apart from vertical pipe runs). For cold and hot services the main differences between our project and the analysed building will be in the distribution runs. These are governed by the number and location of draw-off points. For hot water provision we will assume local water heaters, similar to those provided in the analysed building.

Following on from the discussion concerning sanitary appliances (element 5A), it can be seen that the ratio of draw-off points to gross floor area is not dissimilar in the analysis and our project. However, we know that due to the change in storey heights the lengths of vertical pipe runs must increase. So we will provisionally adopt the figure given in the analysis for the cost of this element per m^2 gfa, increasing it slightly to cover the extra pipework. We will adopt an allowance of £20.50 per m^2.

Update the rate and adjust it for location
$= £20.50 \times 1.15\ (AF) = £23.58$ per m^2

We decide to check that £23.58 per m^2 is a reasonable figure for our project by inspecting several other analyses. We encounter difficulties when we do this for in many analyses this element has been grouped together with several others and no detailed breakdown is available. However, our in-house data sources confirm that our proposed rate per m^2 can be taken as an average figure for office blocks.

Total allowance for element $= £23.58 \times 2246\,m^2 = £52\,961$

	Total	Cost per m^2 gfa
Cost target for water installations	**£52 961**	**£23.58**

Element 5E Heat source, Element 5F Space heating and air treatment

We decide to take these two elements together. The heating and ventilation engineers provide us with an estimate based on the available early design information and the likely heat requirements. This is done by using a combination of approximate quantities and inspection. The engineers estimate the total cost for these two elements at £345 000. This equates to £153.61 per m^2 gfa.

Since the requirements of the new project are similar to the analysed building, we compare this figure with that shown in the analysis. The updated rate in the analysis is £151.17 per m^2 (£131.45 × 1.15). Accordingly we are reassured and adopt their estimate.

	Total	**Cost per m^2 gfa**
Cost target for heat source and space heating and air treatment	**£345 000**	**£153.61**

(*Note:* There would be considerable discussion between the consulting engineer and the cost consultant regarding the formulation of the budget for the element, however an estimate is required. Until the design has been well developed, it is difficult for the engineer to complete the design for the services. It is beyond the scope of this book to deal with this in detail, but we suggest that once you have grasped the fundamental principles of cost planning that you read up further on this topic.)

Element 5H Electrical installations

The information technology revolution, the advent of sophisticated micro-electronic controls for equipment and the need for more electrical power in a building, has meant an increase in the importance of the electrical installation. Finding suitable measures of quantity for all the elements in group 5 can be difficult. Electrical installations is no exception. For example, expressing the costs of this element in terms of cost per m^2 gfa, will not be entirely satisfactory. One method using approximate quantities is to measure the number of outlet points and then apply a rate per point. These rates will change depending on the design and it is usual to state the type of cables and conduits, indicating whether exposed or concealed. This approach has been quite successful with the smaller installations, particularly residential, but less successful with the larger schemes and other building types. We are therefore frequently forced to rely upon inspection for the selection of a cost target. Fortunately cost data exist and can be used to create a range upon which to base our decision.

After discussions with the electrical services engineers we establish that the lighting standard of the building analysed provides suitable levels of illumination for office accommodation. In addition, the provision of power points also appears to match our own requirements. There is a good case for simply updating the cost analysis figure of £57.33 per m^2 gfa. However, before we do this we will add 5% to the rate to give a modest increase to cover extra runs due to increased storey height. Thus

£57.33 × 105% = £60.20

Update the rate and adjust it for location = £60.20 × 1.15 (*AF*) = £69.23

Total element cost = £69.23 × 2246 m^2 = £155 491

	Total	**Cost per m^2 gfa**
Cost target for electrical installations	**£155 491**	**£69.23**

Element 5J Lift installations

There is no lift in the building analysed but for our project the design team, in discussions with the client, have decided that lifts are essential.

This kind of requirement is best met by an enquiry to lift manufacturers and installers. As a result we choose an allowance of £65 000. This is slightly higher than the rate used for the feasibility estimate but the design team have indicated that they wish to raise the specification for the lift cars.

Reduce to cost per m^2:

$$\frac{£65\,000}{2246\,m^2} = £28.94 \text{ per } m^2$$

	Total	Cost per m^2 gfa
Cost target for lift installations	**£65 000**	**£28.94**

Element 5K Protective installations

The requirements are for fire-fighting installations and lightning installations which are best decided by inspection.

We decide on an element allowance of £3.50.

Total element allowance = £3.50 × 2246 m^2 = £7861

	Total	Cost per m^2 gfa
Cost target for protective installations	**£7861**	**£3.50**

Element 5L Communication installations,
Element 5M Special installations

The tenant will be dealing with all items under these elements including security alarms and all installations in connection with information technology. Therefore no allowance is required.

Element 5N Builder's work in connection with services,
Element 5O Builder's profit and attendance on services

A difficult cost target to assess, but best decided by inspection. A number of analyses indicate that the rate included in our analysis is a reasonable sum, but it does not include the work in connection with the lift installation.

We decide to use the figure in the analysis, update it and include the addition for lifts. When we do this we arrive at a rate of £13.21 (£11.79 + £1.42).

Total allowance for element = £13.21 × 2246 m^2 = £29 670

	Total	Cost per m^2 gfa
Cost target for builder's work in connection with services and builder's profit and attendance on services	**£29 670**	**£13.21**

Element 6A Site works

Approximate quantities suggest that an allowance of £118 000 would be realistic. We have not included all the calculation for this element, but it would not be wise to use any other method than measurement.

Reduce to cost per m^2:

$$\frac{£118\,000}{2246\,m^2} = £52.54 \text{ per } m^2$$

	Total	Cost per m^2 gfa
Cost target for site works	**£118 000**	**£52.54**

Element 6B Drainage

Note that this element includes all drainage works outside the building up to and including disposal point, which means connection to the sewer in the public highway.

Adjustment via simple proportion and inspection is rarely appropriate for this element. Therefore, approximate quantities have been used and indicate an allowance of £56 000, following a discussion with the water company who is responsible for the main drainage beyond the site boundary. They have given an approximate depth of the sewer in the road. Some surface-water drainage may be via soakways within the boundary of the site.

Cost per m^2:

$$\frac{£56\,000}{2246\,m^2} = £24.93 \text{ per } m^2$$

	Total	Cost per m^2 gfa
Cost target for drainage	**£56 000**	**£24.93**

Element 6C External services

Approximate quantities suggest that an allowance of £16 000 would be realistic.

Cost per m^2:

$$\frac{£16\,000}{2246\,m^2} = £7.12 \text{ per } m^2$$

	Total	Cost per m^2 gfa
Cost target for external services	**£16 000**	**£7.12**

Comment on preparation of cost plan

There are many differences between our new project and the chosen cost analysis. Therefore, we have had to use approximate quantities to set cost targets for many elements. This was done in full awareness of our previous discussion which stated that adjustment by simple proportion is preferred. Remember, using simple proportion automatically includes allowances for everything one needs for the element in the particular type of building considered. When using approximate quantities, on the other hand, extra care is required for it is easy to overlook details which will only be considered during the detail design stage.

Detailed cost plan

Summary

Element		Allowance: Total cost of element (£)	Allowance: Cost per m² (£)
1	Substructures	55 359	24.65
2A	Frame	155 421	69.20
2B	Upper floors	67 410	30.01
2C	Roof	78 866	35.11
2D	Stairs	62 521	27.84
2E	External walls	102 920	45.82
2F	Windows and external doors	69 225	30.82
2G	Internal walls and partitions	17 404	7.75
2H	Internal doors	11 004	4.90
3A	Wall finishes	44 585	19.85
3B	Floor finishes	134 672	59.96
3C	Ceiling finishes	66 352	29.54
4A	Fittings and furnishings	20 000	8.90
5A	Sanitary appliances	14 360	6.39
5C	Disposal installations	10 669	4.75
5D	Water installations	52 961	23.58
5E 5F	Heat source Space heating and air treatment	345 000	153.61
5H	Electrical installations	155 491	69.23
5J	Lift installations	65 000	28.94
5K	Protective installations	7 861	3.50
5N 5O	Builder's work – services Builder's profit and attendance on services	29 670	13.21
6A	Site works	118 000	52.54
6B	Drainage	56 000	24.93
6C	External services	16 000	7.12
	Sub-total	**1 756 751**	**782.15**
Add	Preliminaries (10%)	175 675	78.22
Add	Contingencies (1.5% approx.)	26 000	11.58
	Sub-total	**1 958 426**	**871.95**
Add	Reserve against unforeseen design eventualities (2.5%)	48 961	21.80
	Reserve against price rises, November 1995 to September 1996 (4.5%)	88 129	39.24
	Grand total	**2 095 516**	**932.99**

As a rough check on our arithmetic,

$$£932.99 \times 2246\,m^2 = £2\,095\,496$$

This is close enough to £2 095 516 for us to rest assured that the discrepancy is due to the approximations made in the conversions to costs per m^2 during the preparation of the cost plan, as noted under element 1, substructures (page 258).

Chapter 11 – Summary

Our feasibility stage estimating, based simply on a statement of office requirements plus quality standards, car parking areas and site conditions, was £2 100 000.

At this stage, our cost plan is based on an expanded brief and the additional information as shown on the scheme design drawings. Changes are apparent in individual element totals but, overall, our cost plan total is more or less the same as our feasibility estimate. This indicates that we should be able to obtain a tender within the cost limit of £2 100 000.

Both the client and the design team have achieved the standards they require without undue restriction. There is no need to consider upgrading any of the allowances in the cost plan. If we suspected that some of the cost targets were too low, we could obviously make additions of up to £4484 at this stage. Further adjustments could be considered via trade-offs between elements, or early use of the design risk reserve.

The design team would, in these circumstances, confirm to the client that the feasibility stage estimate holds good and would go ahead with detailed design in the expectation of receiving a tender at the cost plan figure.

When the cost plan figure exceeds the feasibility and outline proposals stages estimates, the design team must examine the cost plan for all reasonable economies. Only when this has failed will they consider approaching the client for additional funds.

12 Cost Control during Detail Design Stage

This chapter *is* in programmed form and you must follow the page directions exactly.

Turn to page 284

from page 283

The previous stages have all been concerned with *planning* both the design and the cost of the design. The detail design stage is the crucial stage in *controlling* the cost of the detailed design. At the end of this stage, the design team should have completed the detail design of the buildings, in accordance with both the scheme design drawings and the cost plan.

The principles of cost control during detail design were considered in Part 1. Here is a recap of those principles which are incorporated during this stage.

During the detail design stage there must be a method of checking and there must be a means of remedial action. That is, the detailed design of each element should be cost checked as soon as it is completed and remedial action should be taken if necessary.

Cost checking an element consists of estimating the probable cost of the element's detail design. It is generally accepted that the most accurate method of estimating is by approximate quantities based upon a detailed specification. Now that the full specification is available for each element, this accepted traditional technique can be used with confidence.

What prices should generally be used during cost checking?

(a) Prices appropriate to the date of preparation of the first estimate.
(b) Prices appropriate to the date of preparation of the cost plan.
(c) Prices appropriate to the date of cost checking.
(d) Prices appropriate to the date of tender of the proposed building.

Write down an answer from (a) to (d) above.

Turn to page 286

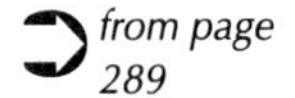
from page 289

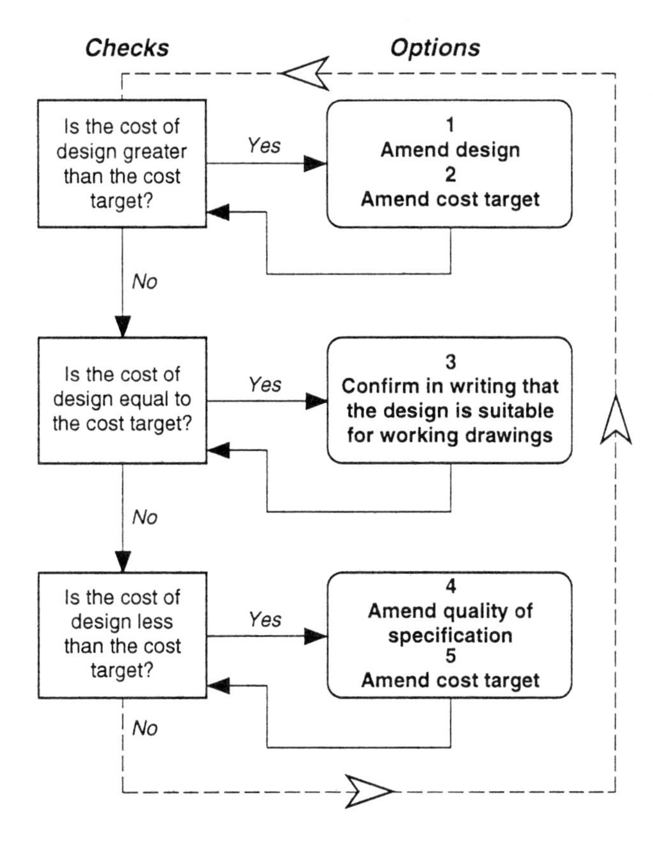

Fig. 12.1

If the outline design and cost plan produced during scheme design are adhered to closely action '3' in Fig. 12.1 should follow the cost checking of most of the elements during detail design.

The usefulness of the cost plan depends for a large part on the cost targets being realistic. Cost targets should therefore only be altered as a last resort, after every alternative design solution for an element has been investigated and rejected as unsuitable. Nevertheless, cost targets may have to be altered as users' requirements become better understood.

If the cost target for an element has to be raised, what else should be done?

Write down your answer.

Turn to page 291

from page 284

(c) Prices appropriate to the date of cost checking.

These are certainly the most appropriate prices to use. They will give accurate cost checks provided you remember to update the element target costs from the cost plan which can be done by using an appropriate cost index.

If you chose

(b) Prices appropriate to the date of preparation of the cost plan.

You are not incorrect, for the prices employed in cost checking would then be appropriate to the date of preparation of the cost plan. However, in this case you would need to convert current costs back to the date of the cost plan. The reason for this is that the approximate quantities estimated from the detail design can only be compared directly with the cost target from the cost plan if the same price level is assumed for both.

Answers **(a)** and **(d)** are both inappropriate for reasons connected with price level and allowance made for design risk.

Write down the possible courses of remedial action which can be taken when cost checking shows that a cost target has been exceeded.

Turn to page 289

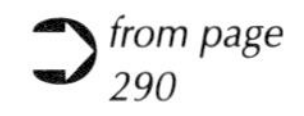
from page 290

Preliminaries, contingencies, and price and design risk, but they must not be ignored.

These allowances cannot be checked in the same way as the others, but they should nevertheless be considered during detail design.

Preliminaries and contingencies can obviously be reviewed as the detailed design progresses, and any necessary changes made.

The price risk part of the price and design risk allowance would have to be changed if an unexpected change in any of the price level factors occurred during the detail design stage. The design risk part of this allowance is used to supply extra money to any of the elements which may run into trouble during the cost checking process.

When all the elements have been designed and all the allowances checked once, what should be done before leaving the detail design stage?

Write down your answer.

Turn to page 292

from page 291

(a) Would be done if any of the previously designed elements were running into trouble at the cost checking stage. If the cost of the design of the other elements were all nearly equal to their cost targets, this action may disturb the balanced design allowed for in the cost plan.

On the other hand, if there are many elements still to be designed and cost checked it may be a good idea to hold on to the savings as an extra 'design risk' allowance.

(b) In fact, all three actions would be 'correct' in different situations, but amending the design is probably the most frequent action taken.

The reason for this is that a cost plan attempts to achieve a balanced design; so increasing the quality of the specification of the element is more likely to give the client good value for money than 'returning' the savings to the client or distributing them among the other elements.

(c) If the design team is happy with the quality of the detail design, it would decide to 'return' the savings to the client.

However, being realistic, most design teams would not choose to do this, unless the detail design of elements is well advanced.

Summarise the basic cycle of operations which should occur during the detail design stage.

Turn to page 290

from page 286

There are two possible courses of remedial action:

(1) the design of the element can be altered in an attempt to bring its cost within the cost target, or,
(2) as a last resort, the cost target for the element can be raised.

The design team has the responsibility of ensuring a response to each cost check. There are three possible scenarios:

(a) the cost of the design can be greater than the cost target,
(b) the cost of the design can be equal to the cost target,
(c) the cost of the design can be less than the cost target.

Bearing this in mind, write down brief descriptions of *all* the possible courses of action which may follow the cost check of an element.

Turn to page 285

from page 288

The basic cycle of operations during the detail design stage is:

(1) Design.
(2) Approximate quantities estimate.
(3) Comparison with cost target.
(4) Decision and action.

This cycle of operations is repeated until all the elements have been designed and cost checked.

Which of the allowances in a cost plan cannot be checked in this way?

Write down your answer.
(If necessary, refer to page 281 to see the detailed cost plan.)

Turn to page 287

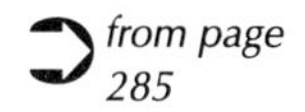
from page 285

In order to avoid exceeding the cost limit, cost targets should be adjusted throughout the cost plan to release surplus funds for the element in trouble.

This highlights another disadvantage of altering a cost target: even if the overall cost limit is not exceeded, the distribution of the available money over the various parts of the building will have to be changed. The result of this may be reduced value for money compared with that envisaged during the preceding stages of the design process.

If the estimated cost of the detailed design of an element turns out to be significantly less than the cost target, what would be the most likely course of action?

(a) The cost targets for the other elements should be amended to absorb the surplus funds.
(b) The quality of the specification of the element should be increased to bring the cost up to the cost target.
(c) The probable savings should be notified to the client.

Write down your answer.

Turn to page 288

from page 287

A final cost check is performed to make sure that nothing has been overlooked.

Notice that the drawings prepared during detail design need not be polished working drawings. These can simply be dimensioned sketches, provided that they contain all the information needed for accurate production drawings which will be prepared before other tender documents (including bills of quantities). Ideally, working drawings should not be prepared for any element until after the final cost check at the end of the detail design stage. It would be unwise to go to the expense of preparing working drawings for a design which may subsequently be changed.

We will now consider two of the refinements to the simplified account of cost checking.

Turn to page 293

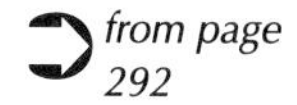
from page 292

The two refinements which we will consider are:

(1) the order in which the elements should be designed and cost checked so as to avoid unnecessary re-designing and re-checking,
(2) the best way to perform cost checking when the detail design is not produced one element at a time.

(1) The order in which elements should be designed

An itemised plan of work and timetable needs to be drawn up at the beginning of the detail design stage. This should set out the intended order in which the elements will be designed. This ensures that time spent during design is as productive as possible.

To draw up an effective cost plan, consideration must be given to the interactive nature of the design and cost planning processes. Imagine the implications where an element such as the frame is found to exceed its target cost, having been designed and cost checked too late in the programme.

There are three characteristics which would make it important to design an element early in the detail design stage. Do you know what they are?

Write down your thoughts.

Turn to page 296

from page 295

(2) The best way to perform cost checking when the detail design is not produced one element at a time

Suppose all the superstructure elements were designed concurrently and completed at the same time. If all the elements were then cost checked, it is possible that one of the important ones, such as the frame, may have overshot its cost target. In this situation all of the superstructure elements may have to be re-designed.

Which of the following ways of dealing with this situation would you adopt?

(a) Always design element-by-element, and plan the order as previously discussed.

(b) Go ahead and re-design all the superstructure elements, hoping that the same situation did not happen too often.

(c) Cost check each element several times during the design of the group by costing the designed part of the element and adding an allowance for the part not yet designed.

Consider your answer.

Turn to page 299

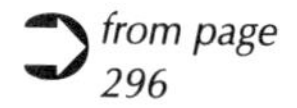
from page 296

Due to the importance of controlling the order in which elements are designed and cost checked, target dates should be set for:

(a) completion of each element's detail design;
(b) completion of each element's cost check;
(c) completion of the action following each cost check.

A lot of information will be circulating among the design team throughout the design process. This will be especially true if remedial action is necessary when any of the elements are cost checked. To avoid confusion between original and amended versions of any item, all cost checks, decisions, and alterations should be clearly communicated.

We will now go on to consider the second point mentioned on page 293.

Turn to page 294

from page 293

The three characteristics which make it desirable to design an element early in the detail design stage are:

(1) Unreliable cost target.
(2) Large cost target (cost-significant element).
(3) A design which affects the design of other elements.

The relative importance of these three characteristics varies from project to project. In general, the second one is less important than the other two (except when it is combined with one of the others).

In addition, the elements which exhibit each of these characteristics will also vary from project to project.

Turn to page 295

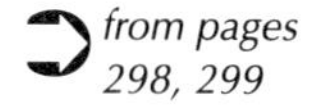
from pages 298, 299

The complications which we have discussed in the last few pages obviously make it essential to have a well-prepared plan of work for the detail design stage. Effective and frequent communication between members of the design team will be required.

Although the detail design stage is the crucial stage for 'cost control during building design', the techniques used are well-established and do not require as much discussion as those used during earlier stages.

A brief summary of the detail design cost control which we have discussed is given on page 300.

If you would like to write your own summary of this chapter, do so now, and compare it with ours.

Turn to page 300

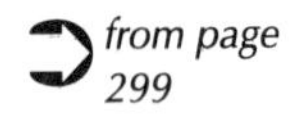
from page 299

Although it may seem rather drawn out, the best solution is to cost check each complicated element several times during its detail design.

Example

Suppose elements A, B, C, and D were being designed concurrently. The first cost check could be performed when approximately a quarter of each element had been designed.

It is easy to prepare an approximate quantities estimate for the designed part of each element. The only slight difficulty arises when allocating allowances for parts of the elements not yet designed. The estimate plus the allowance for each element is then compared with the corresponding cost target, and action is taken.

This process is more involved (and possibly less accurate) than designing and checking one element at a time. Nevertheless it gives valuable information to the design team in time for the designs to be changed without undue trouble.

Turn to page 297

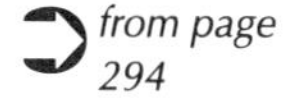
from page 294

(c) Cost check each element several times during the design of the group by costing the designed part of the element and adding an allowance for the part not yet designed.

Although it may be drawn out, the best solution is to cost check each complicated element several times during its detail design.

Answers **(a)** and **(b)** are both inappropriate.

In theory, **(a)** would be the easiest way of avoiding the problem. However, design teams do not design element-by-element; that is not the way the design process works. We must accept that there are often many practical difficulties which make it impossible to produce the detail design of a building in this 'ideal' way.

On the face of it, **(b)** seems to be the easy way out. In practice, however, design teams would find themselves having to do a lot of unnecessary work to regain control over costs if they adopted this method.

The only efficient way of maintaining control is to check at frequent intervals that the frame of reference is being adhered to.

If you would like to see further comment on this point, *Turn to page 298*

If you would prefer to continue, *Turn to page 297* 

from page 297

Summary – Cost control during detail design stage

(1) The latest stage at which each element must be cost checked is as soon as its design is completed.
(2) If several elements are designed concurrently, they should all be cost checked several times during their design.
(3) Whatever the result of a cost check, action must be taken.
(4) There should be a carefully prepared plan of work for the detail design stage.
(5) There should be frequent communications between the members of the design team.
(6) All communications should be recorded and dated.

You have now completed Part 2 of this programme.

To revise this part of the programme, answer the questions on page 301. The answers are given on pages 302–304.

Turn to page 301

Part 2 Test

1 What is an elemental cost analysis?

2 Briefly describe four uses of elemental cost analyses.

3 What yardstick of cost is used in cost analyses?

4 List the major factors which affect the costs of elements, and discuss each of them briefly.

5 Define 'element unit rate', and give an example.

6 Explain how a first estimate is prepared by the interpolation method.

7 (a) Describe the cost allocations and allowances which are included in an outline cost plan.

(b) Explain how they are prepared.

8 How does the cost planning performed during outline proposals differ from the cost planning during (a) feasibility, and (b) scheme design?

9 Name and describe three methods of preparing the cost target for an element.

10 Prepare a cost target for internal doors on the basis of the following information:

Element unit rate in the analysis = £145 per m^2
Price level index at analysis tender date = 110
'Current' price level index = 120
Total floor area of proposed building = 2500 m^2
Area of external doors in new project = 112 m^2

It is decided to deduct 10% to allow for a reduction in the quality of the external doors in the new project.

11 Describe the possible events which may follow the cost checking of an element.

12 Imagine a project with a cost limit of £750.00 per m^2. When half of the elements have been cost checked, the cost of the design is running at £11 per m^2 greater than the sum of the cost targets concerned. The cost check of the next element shows that it is £15 per m^2 over its cost target. What should be done?

13 List the considerations which should be borne in mind when deciding the order in which the detailed designs of the elements should be produced and checked.

Answers overleaf 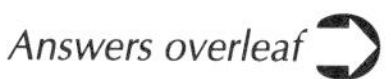

Answers to Part 2 Test

Marks

1 An elemental cost analysis is the analysis of the client's costs (1)
on a tender to determine the cost of each element of (2)
a building.
3

2 *Appreciation:* to enable clients and designers to appreciate
how cost is distributed among the elements of a building. (1)
Judgement: to enable clients and designers to develop ideas (1)
as to how cost could have been allocated to obtain a
more balanced design. (1)
Belated remedial action: to allow remedial action to be taken on
receipt of high tenders by revealing the sources of
over-expenditure. (1)
Planning: to help with the cost planning of future projects. (1)
5

3 Cost per m^2 of gross floor area. **2**

4 *Quantity:* the amount of each element in the building; (1)
this is affected by both floor area and the 'density' of each
element. (1)
Quality: the quality of the materials used for an element (1)
and the relative proportions of the various qualities of
specification within the element. (1)
Price level: this covers general market price level, contractor's (1)
price level, site conditions, contract conditions, etc. (1)
6

5 $$\text{Element unit rate} = \frac{\text{cost of element}}{\text{quantity of element}}$$ (2)

e.g. element unit rate for internal doors = £131.27 per m^2. (1)
3

6 (a) A range of cost analyses of the same *type* of building
as the proposed building is assembled. (1)
(b) The cost analyses of the buildings whose *quality* (1)
approximates most closely to that desired by the client (1)
are selected from this range, (between 4 and 10). They
are then adjusted for time and location. (1)

Turn to page 303

Marks

(c) The major differences between the chosen analyses and (1)
the new project are *isolated* by comparing the initial brief
of the proposed building with all of the information
given in the analyses. (1)
(d) Allowances are made for each of these differences (1)
(including floor area, site works, etc.). (1)
(e) Additions are made for preliminaries and contingencies. (1)
Then a reserve against design risk is included. (1)
(f) An allowance is made as a reserve against price rises
between the preparation of the first estimate and the
contractor's actual price level on the tender. (1)

11

7 (a) An allowance is made for 'price and design risk' (1)
by assessing the cost of unforeseen design difficulties and (1)
possible price rises between the preparation of the outline
cost plan and the receipt of tender. (1)
(b) Cost targets are prepared for each group of elements by
allowing for the major differences between the buildings (1)
analysed and the new project, then updating to the date (1)
of preparation of the outline cost plan. (The major (1)
differences are allowed for by comparing all the
information available for both the new project and the
cost analyses.)

6

8 The differences which are isolated and allowed for are more
detailed than in feasibility, but less detailed than in scheme (1)
design. The available money is allocated to the various parts (1)
of the building during outline proposals, whereas no (1)
allocation takes place at feasibility: this allocation is not as (1)
detailed as in scheme design.

4

9 *Simple proportion:* The element cost selected from the cost (1)
analyses chosen at feasibility is adjusted by proportion (1)
for Price, Quantity, and Quality. (1)
Inspection: The target cost per m^2 of floor is chosen from a (1)
range obtained from a selection of suitable cost analyses (2)
and/or cost studies. (1)
Approximate quantities: A provisional specification of the
required quality is prepared for the element. Measurement (2)
and pricing then proceeds using the approximate quantities (2)
technique. **11**

Turn to page 304 ➲

Marks

10 *Adjustment for Price*

$$\text{'Current' element unit rate} = \frac{120}{110} \times £145.00 \text{ per m}^2$$
$$= £158.18 \text{ per m}^2 \quad (2)$$

Adjustment for Quantity

$112\,\text{m}^2 \times £158.18 = £17\,716$ (2)

$$\text{Reduce to cost per m}^2 = \frac{£17\,716.16}{2500\,\text{m}^2} = £7.09 \text{ per m}^2 \quad (2)$$

Adjustment for Quality

90% of £17 716 = £15 944

90% of £7.09 per m^2 = £6.38 per m^2 (3)

	Total	Cost per m^2
Cost target for internal doors	£15 944	£6.38

9

11 (a) If the cost of the design equals the cost target, the design is confirmed in writing as being suitable for working drawings. (2)

(b) If the cost of the design exceeds the cost target, the design can be changed or, if this is impossible, the cost target can be increased by borrowing from the other (2) elements (1)

(c) If the cost of the design is less than the cost target,
the quality of the design can be increased, (1)
the savings can be distributed among the other elements, (1)
or the savings can be left as a saving for the client. (1)

8

12 Re-design, if possible, to come within cost target. **2**

13 (a) How much the design of the element affects the design of other elements. (2)

(b) How unreliable the cost target is known to be (2)

(c) How large the cost target is relative to the others. (1)

5

TOTAL **75**

Appendix A

Detailed Cost Analysis and Element Costs

BCIS On-line Analysis No. xxx

CI/SfB 320
Offices
BCIS Code: A-2-1408

DETAILED COST ANALYSIS

Job Title:	Office Units, Red Knights		
Location:	Reading, Berkshire		
Client:	Unipol Properties Ltd.		
Date for receipt:	10 September 1992	Date of tender:	September 1992

INFORMATION ON TOTAL PROJECT

Project details:
2 storey office block with self contained units on 2 floors with plant rooms in the roof voids together with external works including car parking, fencing, landscaping, services and drainage.

Site conditions:
Level demolition site with bad ground conditions but above water table. Unrestricted working space and access. Contaminated ground, approximately 400 mm excavate and removed. All excavated material removed to licensed tip.

Market conditions:
Generally competitive overall but not keen. Lowest tender not accepted due to unacceptable qualification.

			Competitive Tender List
Project tender price index	106 (base: 1985 BCIS Index Base)		
Tender documentation:	Bill of Quantities		£989,000 *
Selection of contractor:	Selected competition		£1,167,043
Number of tenders:	issued 6		£1,592,203
	received 6		£1,901,202
			£2,118,457
Type of contract:	JCT private contract 1980 edition		
Cost fluctuations:	Firm		
Contract period	stipulated by client	9 months	
	offered by builder	9 months	
	agreed	9 months	

ANALYSIS OF SINGLE BUILDING

Accommodation and design features:

2 storey office building with 2 self contained units on 2 floors with plant rooms in the roof voids. Concrete stanchion bases and ground slab.
Steel frame. Lightweight reinforced concrete upper floors on steel decking. Steel and timber roof with fibre cement slates. Facing brick/block cavity walls. Specialist curtain walls and external windows. Plasterboard ceilings and raised access floors. Air conditioning, electric installations, lightning protection.

Areas:

Basement floors	-
Ground floor	704 m^2
Upper floors	704 m^2
Gross floor area	1,408 m^2
Usable floor area	1,195 m^2
Circulation area	75 m^2
Ancillary area	74 m^2
Internal divisions	64 m^2
Gross floor area	1,408 m^2

Functional units

1,195 m^2 usable floor area

Percentage of gross floor area
2 storey construction 100%

Floor space not enclosed	3,872 m^3
Internal cube	835 m^2
Wall to floor ratio	0.59

Storey heights:

Average below ground floor	-
at ground floor	2.75 m
above ground floor	2.75 m

BRIEF COST INFORMATION

TOTAL CONTRACT

Measured work	845,192	
Provisional sums	58,500	
Prime cost sums	150,000	
Preliminaries	100,101	being 9.5% of remainder of contract sums (less contingencies)
Contingencies	13,250	
Contract sum	£1,167,043	

* Lowest tender not accepted, see MC text

Element costs | CI/SfB 320 | Offices

Gross internal floor area: 1408 m^2 | Date of tender: 10 September 1992

	Element	Preliminaries shown separately				Preliminaries apportioned	
		Total cost of element (£)	Cost per m^2 (£)	Element unit quantity	Element unit rate	Total cost of element (£)	Cost per m^2 (£)
1	SUBSTRUCTURES	45,243	32.13	704 m^2	64.27	49,541	35.19
2A	Frame	45,366	32.22	1408 m^2	32.22	49,676	35.28
2B	Upper floors	22,629	16.07	704 m^2	32.14	24,779	17.60
2C	Roof	64,491	45.80	900 m^2	71.66	70,618	50.15
2D	Stairs	15,758	11.19			17,255	12.25
2E	External walls	27,069	19.23	304 m^2	89.04	29,641	21.05
2F	Windows and external doors	172,555	122.55	531 m^2	324.96	188,948	134.20
2G	Internal walls and partitions	26,295	18.68			28,793	20.45
2H	Internal doors	8,270	5.87	63 m^2	131.27	9,056	6.43
	SUPERSTRUCTURES	382,433	271.61			418,764	297.42
3A	Wall finishes	18,930	13.44	904 m^2	20.94	20,728	14.72
3B	Floor finishes	64,650	45.92	1301 m^2	49.69	70,792	50.28
3C	Ceiling finishes	19,814	14.07	1380 m^2	14.36	21,696	15.41
	INTERNAL FINISHES	103,394	73.43			113,216	80.41
4A	Fittings and furnishings	10,195	7.24			11,164	7.93
5A	Sanitary appliances	8,110	5.76			8,880	6.31
5B	Services equipment						
5C	Disposal installations	5,477	3.89			5,997	4.26
5D	Water installations	26,921	19.12	56 nr. draw off points	480.73	29,478	20.94

5E	Heat source				
5F	Space heating & air treatment	185,081	131.45	202,664	143.94
5G	Ventilating systems				
5H	Electrical installations	80,721	57.33	88,389	62.78
5I	Gas installations				
5J	Lift installations				
5K	Protective installations				
5L	Communications installations				
5M	Special installations				
5N	Builder's work in connection	14,432	10.25	15,803	11.22
5O	Builder's profit & attendance				
	SERVICES	320,742	227.80	351,212	249.44
	BUILDING SUB-TOTAL	862,007	612.22	943,898	670.38
6A	Site works	141,532	100.52	154,978	110.07
6B	Drainage	41,816	29.70	45,789	32.52
6C	External services	8,337	5.92	9,129	6.48
6D	Minor building works				
	EXTERNAL WORKS	191,685	136.14	209,895	149.07
7	Preliminaries	100,101	71.09		
	TOTAL (less contingencies)	1,153,793	819.46	1,153,793	819.46

	Specification and design notes				CI/SfB: 320 Offices September 1992
			Unit quantity	All-in unit rate	Cost (£)
1	SUBSTRUCTURES	**Total**			45,243
	Plain concrete Grade 25 in stanchion bases, and strip foundations *Reinforced concrete Grade 30 in beds 150-300 mm thick*		704 m^2	64.27	
2A	Frame	**Total**			45,366
	Traditional steel frame, galvanised 'Z' roof purlins, beams for plant rooms in roof space *Fire protective coating to first floor beams, fire protective casings to exposed columns*				
2B	Upper floors	**Total**			22,629
	Lightweight concrete reinforced with fabric on proprietary ribbed steel decking. *Reinforced in situ concrete floors to plant rooms* *Concrete casing to steel beams.*				
2C	Roof	**Total**			64,491
	Artificial slates on battens on insulated metal decking over roof purlins to sloping roof.				
2D	Stairs	**Total**			15,758
	Main staircase of concrete with carpet finish to treads and risers, tubular steel open balustrades, PVC sheathed handrails.		2 Nr	4,253	8,506
	External escape stairs, galvanised steel spiral pattern.		2 Nr	3,626	7,252
2E	External walls	**Total**			27,069
	Decorative brick casing to steel columns forming piers between window bays.		107 m^2	108.98	11,661
	Bricked face hollow walls up to window sill level to both floors.		197 m^2	78.21	15,408

2F	Windows and external doors	**Total**			172,555
	Colour coated aluminium framed curtain walling with top-hung opening lights, tinted sealed double glazing units.		507 m^2	329.49	167,052
	Fire escape door, loading doors and pair of main entrance doors to each unit.		24 m^2	229.29	5,503
2G	Internal walls and partitions	**Total**			26,295
	100 mm Blockwork division walls		162 m^2	19.80	3,208
	200 mm Blockwork division walls		116 m^2	30.12	3,494
	One-brick walls		123 m^2	45.15	5,553
	Demountable partitions		112 m^2	125.36	14,040
2H	Internal doors	**Total**			8,270
	Teak veneered flush doors, generally half or one-hour fire resistance, stained hardwood frames, linings and architraves Satin anodised aluminium ironmongery		18 sgles 10 dbles	240 395	4,320 3,950
3A	Wall finishes	**Total**			18,930
	Two-coat plaster to block walls		720 m^2	14.20	10,224
	Ceramic tiles full height toilet areas and splashbacks in kitchens		184 m^2	47.32	8,706
3B	Floor finishes	**Total**			64,650
	Office areas: Pedestal access floors, hardwood skirtings. Stain and matt finish to skirtings. Lay only carpet tiles (supplied by client).		1,040 m^2	55.51	57,730
	Toilets and other areas: cement and sand screed, vinyl tiles and coved skirting.		261 m^2	26.51	6,920
3C	Ceiling finishes	**Total**			19,814
	Plasterboard and skim coat on timber framing. Emulsion paint.		1,380 m^2	14.36	

4A	Fittings and furnishings	**Total**			10,195
	Sundry fittings in entrance lobbies, including pinboards etc.				1,400
	Melamine faced chipboard vanitory tops in toilets				1,600
	Melamine faced chipboard duct cladding				795
	Grab rails etc. in disabled toilets				1,200
	Louvre blinds				5,200
5A	Sanitary appliances	**Total**			8,110
	Urinals		6 Nr	171	1,026
	Wash hand basins		14 Nr	171	2,394
	WC suites		14 Nr	190	2,660
	Cleaners' sinks		4 Nr	210	840
	Short connections to services)				
	PVC overflows. Sundries)				1,190
5C	Disposal installations	**Total**			5,477
	PVC waste pipes and soil and ventilation pipes				
5D	Water installations	**Total**			26,921
	Mains supply				
	Cold water service				
	Hot water by electric water heaters		56 draw off points	375.38	21,021
5E	Heat source	**Total**			
	Included in element 5F and 5D				
5F	Space heating & air treatment	**Total**			185,081
	Four-pipe fan coil comfort cooling system				
	Toilet extract ventilation				
5G	Ventilating systems	**Total**			
	Included in element 5F.				

5H	Electrical installations	**Total**			80,721
	Office and general lighting. *Emergency and external lighting.* *Fire alarms.* *Office and general power supplies.* *Lightning protection.*				
5N	Builder's work in connection	**Total**			14,432
	Normal builder's work in connection with service installations.				
6A	Site works	**Total**			141,532
	Macadam-surfaced car parking area. *PCC block paving to pedestrian access.* *Landscaping.*				
6B	Drainage	**Total**			41,816
	Soil and surface water drainage systems in plastic or glazed clay ware pipes and fittings; granular beds and surrounds. *Precast concrete and brick manholes.*				
6C	External services	**Total**			8,337
	Builder's work only for water and electrical services (no cost allowed for service connection/supplies by local utilities authorities). *Builder's work for telephone and lightning protection.*				
7	Preliminaries	**Total**			100,101
	9.5% of remainder of Contract Sum (exc. Contingencies).				

Proposed two-storey office block – ground floor plan

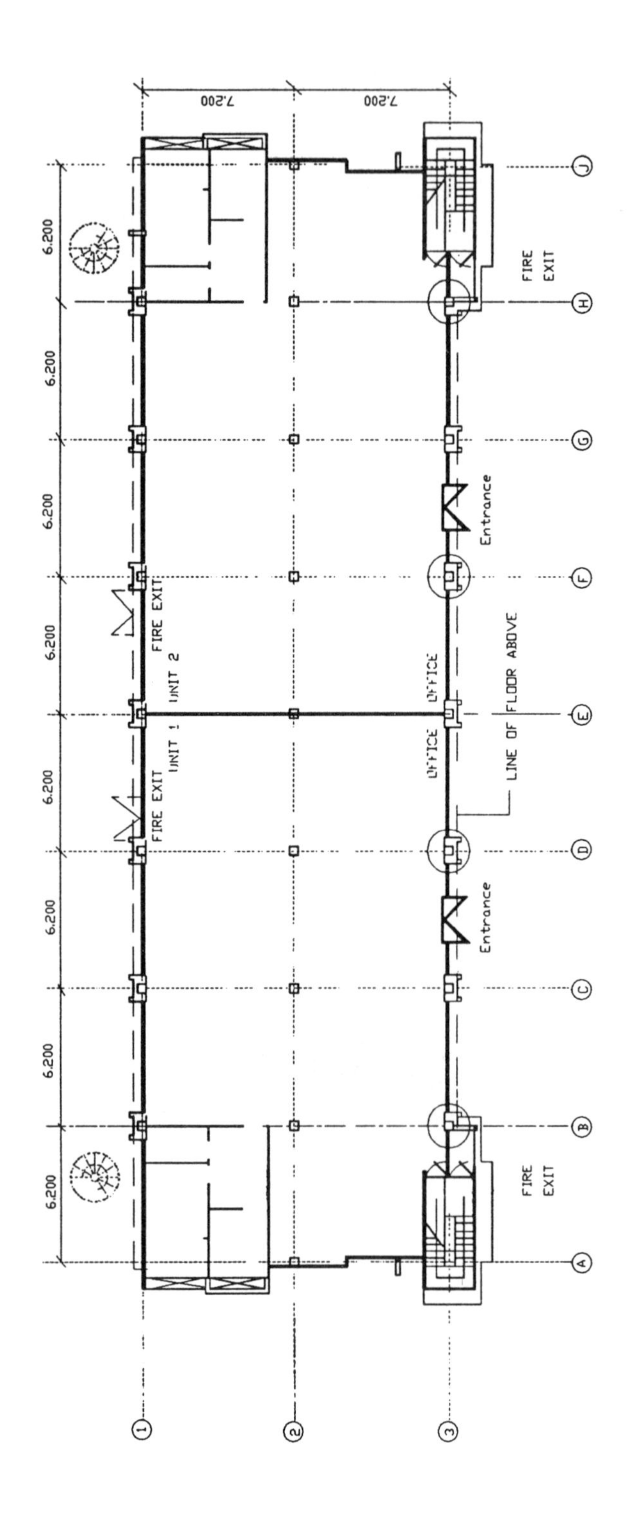

Proposed two-storey office block – elevations

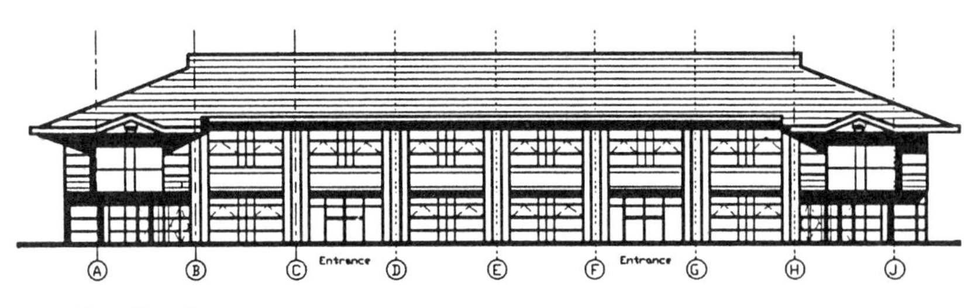

Front Elevation

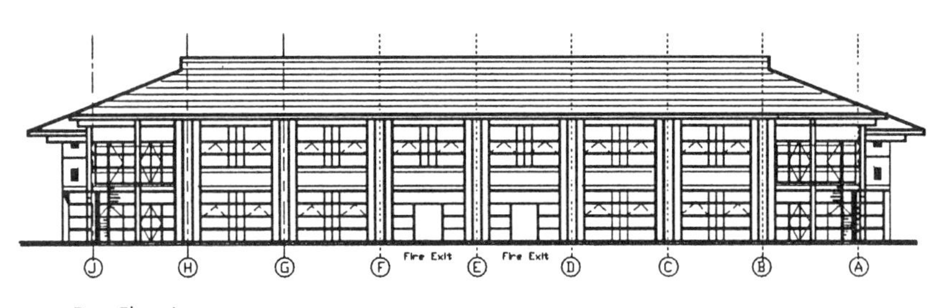

Rear Elevation

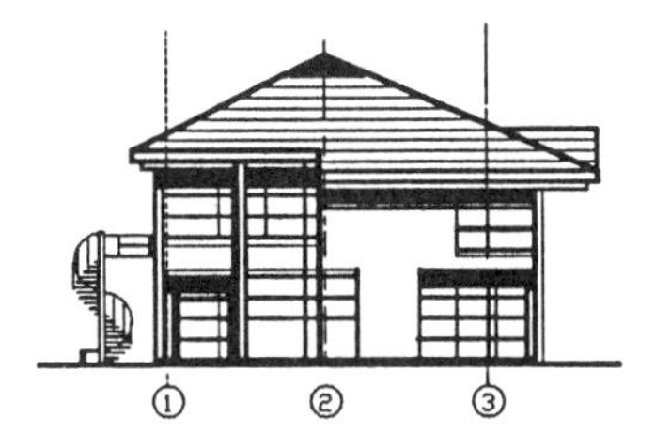

Side Elevation (handed)

Appendix B

Plans and Elevations for Proposed Office Block and Design Process Terminology

Proposed three-storey office block – ground floor plan

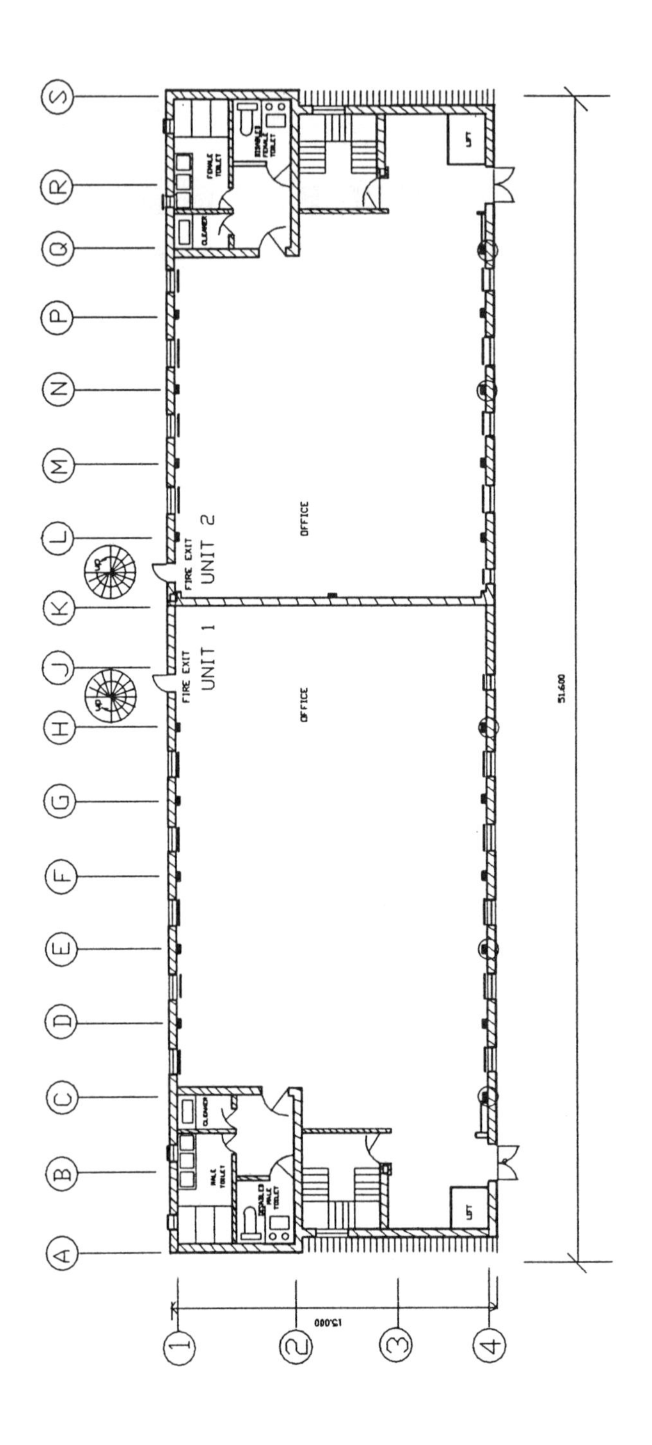

Proposed three-storey office block – elevations

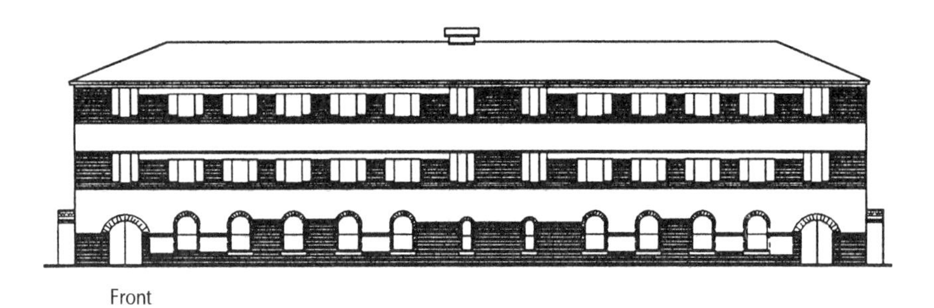

Front

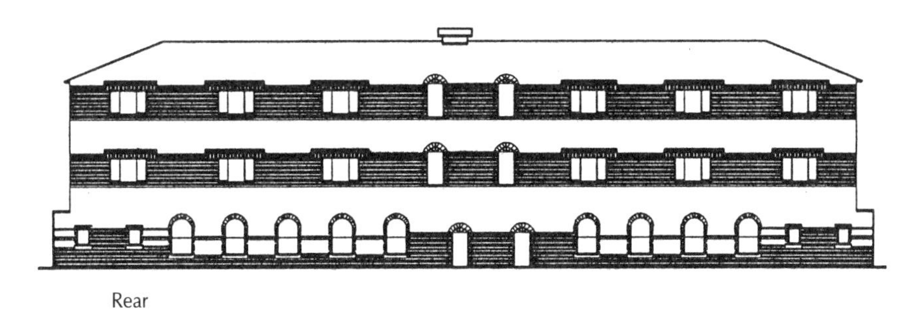

Rear

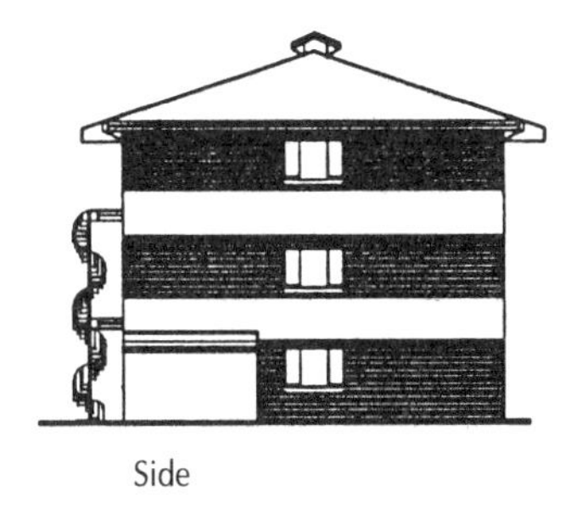

Side

Design process terminology

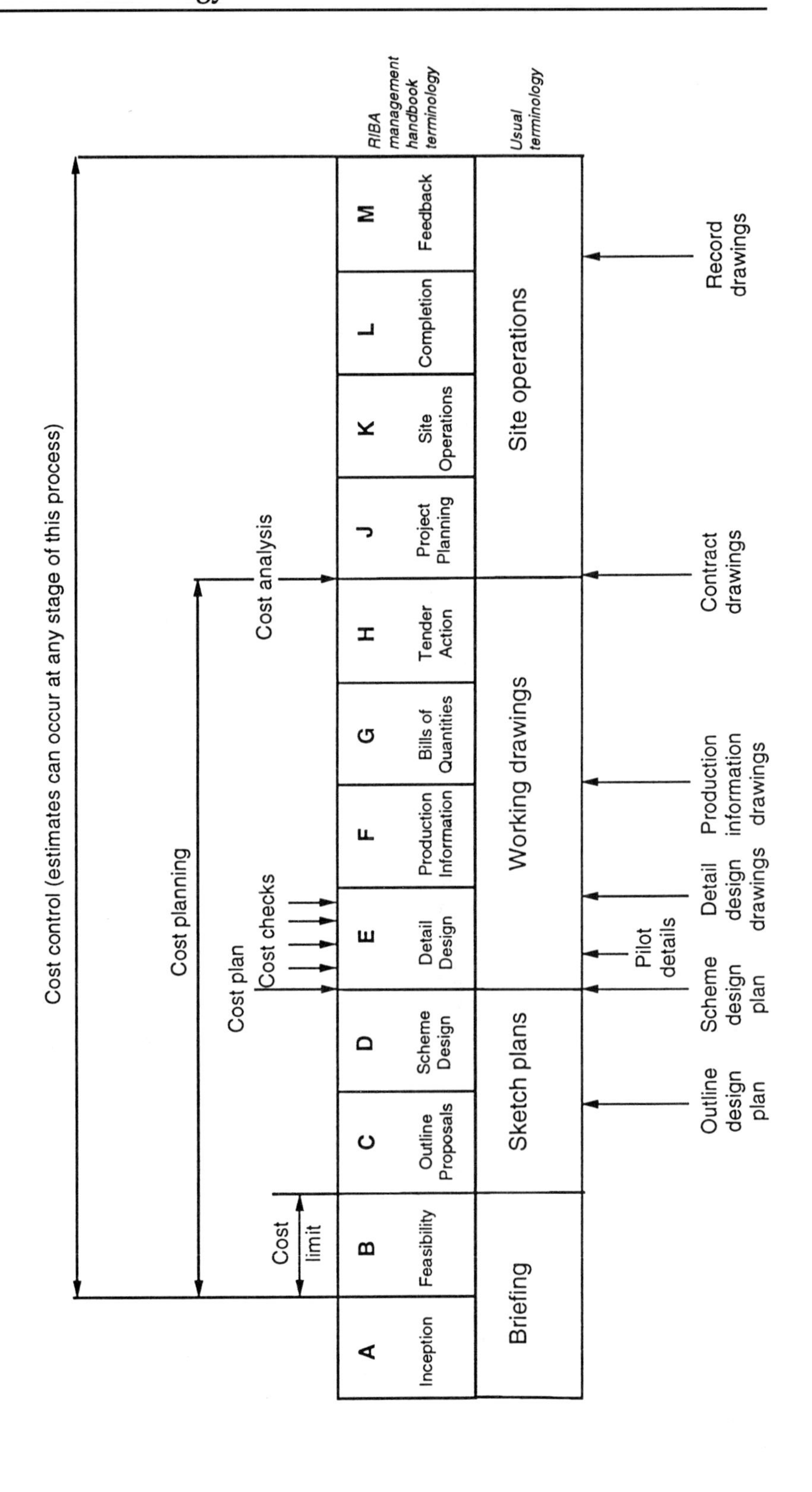

Index